THE TERRY LECTURES

NATURAL REFLECTIONS

Other Volumes in the Terry Lectures Series Available from Yale University Press

The Courage to Be Paul Tillich
Psychoanalysis and Religion Erich Fromm
Becoming Gordon W. Allport
A Common Faith John Dewey
Education at the Crossroads Jacques Maritain
Psychology and Religion Carl G. Jung
Freud and Philosophy Paul Ricoeur
Freud and the Problem of God Hans Küng
Master Control Genes in Development and Evolution Walter J. Gehring
Belief in God in an Age of Science John Polkinghorne
Israelis and the Jewish Tradition David Hartman
The Empirical Stance Bas C. van Fraassen
One World Peter Singer
Exorcism and Enlightenment H. C. Erik Midelfort
Reason, Faith, and Revolution Terry Eagleton
Thinking in Circles: An Essay on Ring Composition Mary Douglas
The Religion and Science Debate: Why Does It Continue? Edited by Harold W. Attridge

Natural Reflections

Human Cognition at the Nexus of Science and Religion

BARBARA HERRNSTEIN SMITH

Yale University Press New Haven and London

Set in Janson type by Tseng Information Systems, Inc., Durham, North Carolina.
Printed in the United States of America.

Library of Congress Cataloging-in-Publication Data
Smith, Barbara Herrnstein.
Natural reflections : human cognition at the nexus of science and religion / Barbara Herrnstein Smith.
p. cm. — (The Terry lectures)
Book is adapted from the Dwight H. Terry Lectures delivered at Yale University in 2006.
Includes bibliographical references and index.
ISBN 978-0-300-14034-7 (cloth : alk. paper)
1. Religion and science. 2. Cognition. I. Title.
BL241.S625 2009
201'.65—dc22

2009023111

A catalogue record for this book is available from the British Library.

This paper meets the requirements of ANSI/NISO Z39.48-1992 (Permanence of Paper).

10 9 8 7 6 5 4 3 2 1

The Dwight Harrington Terry Foundation Lectures on Religion in the Light of Science and Philosophy

The deed of gift declares that "the object of this foundation is not the promotion of scientific investigation and discovery, but rather the assimilation and interpretation of that which has been or shall be hereafter discovered, and its application to human welfare, especially by the building of the truths of science and philosophy into the structure of a broadened and purified religion. The founder believes that such a religion will greatly stimulate intelligent effort for the improvement of human conditions and the advancement of the race in strength and excellence of character. To this end it is desired that a series of lectures be given by men eminent in their respective departments, on ethics, the history of civilization and religion, biblical research, all sciences and branches of knowledge which have an important bearing on the subject, all the great laws of nature, especially of evolution . . . also such interpretations of literature and sociology as are in accord with the spirit of this foundation, to the end that the Christian spirit may be nurtured in the fullest light of the world's knowledge and that mankind may be helped to attain its highest possible welfare and happiness upon this earth." The present work constitutes the latest volume published on this foundation.

For Irwin

Contents

Preface xi

Acknowledgments xvii

CHAPTER ONE: Introduction: Prophecies, Predictions, and Human Cognition 1

CHAPTER TWO: Cognitive Machinery and Explanatory Ambitions: The New Naturalism 25

CHAPTER THREE: "The Gods Seem Here to Stay": Naturalism, Rationalism, and the Persistence of Belief 59

CHAPTER FOUR: Deep Reading: The New Natural Theology 95

CHAPTER FIVE: Reflections: Science and Religion, Natural and Unnatural 121

Notes 151

Works Cited 179

Index 195

Preface

The present book, adapted from the Dwight H. Terry Lectures delivered at Yale University in 2006, examines a set of contemporary intellectual projects involving the relations among science, religion, and human cognition. These include recent studies seeking to explain religion on the basis of cognitive science and evolutionary theory, writings by theologians seeking to demonstrate the cognitive compatibility of natural science and traditional religious teachings, and efforts by scientists and philosophers to distinguish sharply between religious belief and scientific knowledge on the basis of their supposed respective cognitive features.

In examining these projects, I draw on views of belief, cognition, and scientific knowledge elaborated in two previous works, *Belief and Resistance: Dynamics of Contemporary Intellectual Controversy* (Cambridge, MA: Harvard University Press, 1997) and *Scandalous Knowledge: Science, Truth and the Human* (Edinburgh: Edinburgh University Press, 2005/ Durham, NC: Duke University Press, 2006). These views, notably an ecological-dynamic conception of cognition and a constructivist-pragmatist account of the formation and stabilization of knowledge, are at some variance with mainstream cognitive science and traditional rationalist epistemology (see chapter 1, "Introduction: Prophecies, Predictions, and Human Cognition"). As I hope this book demonstrates, however, these alternative conceptions and accounts offer useful perspectives on a number of phenomena and issues central to contemporary thinking about science and religion. In particular, conceptions of cognition that acknowledge the significance of ongoing organism-environmental interactions and stress both the motility and embodiment of cognitive processes illuminate crucial features of belief that remain elusive or invisible under the prevailing mentalist, innatist, computational-modular models. Similarly, constructivist and pragmatist understandings of knowledge, along with the accounts of scientific practice developed in contemporary history and sociology of science, alert us to resemblances, connections, and continuities between science and religion where their relations are otherwise represented primarily in terms of difference, disjunction, and opposition.

Of particular interest here is a set of recently published self-consciously innovative works, largely by anthropologists and psychologists, offering to explain religion on the basis of evolutionary theory and cognitive science—a programmatic effort that I call the New Naturalism (see chapter 2, "Cognitive Machinery and Explanatory Ambitions: The New Naturalism"). Noting the strenuous skepticism, theological and other, regarding the adequacy or even possibility of such explanations, I draw attention to a long naturalistic tradition in the study of religion and observe the intellectual promise of approaches informed by biology, cognitive neuroscience, and related fields. I go on to suggest, however, that New Naturalist accounts of religion are currently limited by dubious conceptual and methodological commitments and related intellectual biases, many of them taken over from paradigmatic evolutionary psychology (see chapter 3, "'The Gods Seem Here to Stay': Naturalism, Rationalism, and the Persistence of Belief"). The intellectually confining consequences of these commitments are detailed mainly through discussions of two influential New Naturalist books, Pascal Boyer, *Religion Explained: The Evolutionary Origins of Religious Thought* (New York: Basic Books, 2001), and Scott Atran, *In Gods We Trust: The Evolutionary Landscape of Religion* (Oxford: Oxford University Press, 2002). The more general characteristics and limits of current cognitive-evolutionary approaches are indicated through comparison with other naturalistic accounts of religion, both old and new. Especially useful in this regard is a somewhat earlier study, *Creation of the Sacred: Tracks of Biology in Early Religions* (Cambridge, MA: Harvard University Press, 1995), by the German classicist Walter Burkert. Like the New Naturalists, Burkert draws centrally on evolutionary biology and such related fields as game theory and primatology, but his accounts of the origins and early forms of religious ideas and practices illustrate the value of more historically grounded and intellectually spacious approaches.

A second project of particular interest here, which I call the New Natural Theology, is the attempt, largely by theologians but also by some theistic scientists, to reveal the compatibility of traditional religious beliefs and current scientific accounts of the natural world. Such efforts include a widely cited work by theologian John Haught titled *Deeper Than Darwin: The Prospect for Religion in the Age of Evolution* (Boulder, CO:

Westview Press, 2003). In chapter 4, "Deep Reading: The New Natural Theology," I consider some characteristic conceptual and rhetorical features of the New Natural Theology as illustrated in Haught's allegorical reading of what he sees as the narrative of biological evolution. Also of interest here is Haught's related critique of current evolutionary explanations of religion. Although his charge of self-refutation against such explanations misses its mark, many New Naturalists, in implicitly or explicitly exempting their own cognitive activities and intellectual enterprises from naturalistic explanation, are vulnerable to the theist's classic taunt of "*tu quoque*": you, too—*you* do just what you say *we* do.

The two intellectual projects described above, one a naturalizing of religion, the other a theologizing of nature, could be seen as mirror images of each other—"natural reflections" in that sense. But the reflections indicated by my title have broader reference points. They include the ways in which these diverse and sometimes mutually opposed intellectual projects reflect more general and often mutually illuminating features of human cognition and also the complex and sometimes curious operations, in all these debates, of such key terms and concepts as "natural," "naturalism," and "nature" itself.

A number of recent works proposing strong and often polemically framed contrasts between science and religion—and, in some cases, between science and all other intellectual enterprises—are discussed in chapter 5, "Reflections: Science and Religion, Natural and Unnatural." Especially notable here is the idea that, cognitively and evolutionarily, religion is "natural" while science is "unnatural." As I indicate, the arguments central to this idea require strained definitions of both religion and science along with the obliteration of quite a bit of recorded human history. I go on to discuss the recurrence of related Two Cultures caricatures—science versus the humanities—in current promotions of New Naturalist accounts of religion, illustrated here by Daniel Dennett's *Breaking the Spell: Religion as a Natural Phenomenon* (New York: Viking, 2006).

In the concluding section of the book, drawing together a number of themes and earlier observations, I frame a view of the relation between science and religion that recognizes the distinctive epistemic value of science but does not yield a dubious science-exceptionalism with regard to the more general intellectual efforts and cognitive achievements and liabilities of human beings.

This book is the product of a four-year-long foray into the archives of religious studies and theology, where, entering as a relative novice, one soon discovers that almost anything one might be inclined to say in these realms, whether wise or foolish, has already been said innumerable times before, often for several centuries and sometimes for two or three millennia. Nevertheless, ancient wisdom must be made responsive to contemporary concerns, ancient folly must be challenged in its modern forms, and, as one may also discover from such forays, it's a good idea to pause from time to time to be sure one knows which is which.

Acknowledgments

A number of people must be thanked for enabling the intellectual adventure represented by this book, beginning with the members of the Terry Committee for inviting me—and especially Dale Martin for persuading me—to deliver the lectures on which it is based. Special thanks to Harry Attridge, William Summers, and Leo Hickey for their hospitality during my stay at Yale and to Lauralee Field for her managerial attentions. My thanks as well to Drew McDermott, Robert Shulman, and Timothy Goldsmith for genial and instructive exchanges.

Friends, colleagues, and students at Duke University set me in good directions and helped along the way: Bruce Lawrence and Elizabeth Clark provided valuable introductions to the field of religious studies; Robert Brandon, Güven Güzeldere, Andrew Janiak, and Grant Ramsey were reliable sources for good tips and lively conversations about cognitive science, the history and philosophy of science, and the philosophy of biology; Roy Weintraub and Julie Tetel were attentive readers of early drafts; Claudia Koonz and Srinivas Aravamudan provided worldly news and cosmopolitan perspectives.

My thanks to the students in the "Natural and Supernatural" seminars at Duke and Brown for their spirited responses to early versions of many of these ideas and to Nancy Armstrong and Leonard Tennenhouse for giving encouraging ears and eyes to the Terry Lectures. I am especially grateful to David Rubin and Kenneth Caneva for extra-sharp readings of several chapters and to Susan Oyama for exceptionally useful comments on a first draft of the lectures and for ongoing illuminations of the toughest questions. Readers for Yale University Press offered valuable responses and suggestions, all genuinely appreciated if not all heeded.

My warm thanks to Fenja Brodo, Julia Weitzman-Manganaro, and Deirdre Smith for sharing their experiences and insights, and to Stephen Barber, whose subtle and companiable views on perennial issues have helped me to better assess and articulate my own.

This book is dedicated to my brother, Irwin Brodo, who has figured for me for the past fifty years as an exemplary scientist and forever as an exemplary human being.

CHAPTER ONE

Introduction: Prophecies, Predictions, and Human Cognition

This book is concerned in large part with the processes of human cognition, both how they are theorized and how they appear to operate at some significant contemporary junctures of religion and science. To introduce some of the themes associated with this topic, I relate below the story of a famous study in social psychology. Of particular interest is the duplicative behavior of the two sets of people involved: one, a team of psychologists testing a new hypothesis about human cognition; the other, a group of quasi-millenarians—or, as we might say now, End Timers—whose behavior the psychologists were observing. In commenting on the story, I suggest that the scientists and their subjects shared certain cognitive tendencies that could be seen as characteristic of humans generally. I go on to outline a set of contemporary views of cognition that I think illuminate those tendencies along with a number of related features of human behavior, both religious and scientific, that will concern us in later chapters.

A NATURAL EXPERIMENT

I begin with the story.

One September evening in the early 1950s, Marion Keech, resident of a small midwest town in the United States, phoned her local newspaper to report an important message she had received via automatic writing from a superior being on another planet. As Keech explained, the author of the message had visited the earth on a flying saucer and observed a number of fault lines in the earth's crust that foretold an impending large-scale flood. The flood would begin just before dawn on December 21 and spread inland from the local great lake to the Gulf of Mexico. It would be accompanied by a cataclysm on the west coast, extending from Seattle to South America, and ultimately destroy the entire hemisphere. The following day, the *Lake City Herald* printed a report of Keech's phone call under the headline: "PROPHECY FROM PLANET. CLARION CALL TO CITY: FLEE THAT FLOOD. IT'LL SWAMP US ON DEC. 21, OUTER SPACE TELLS SUBURBANITE."[1]

A few weeks later two members of a team of social psychologists at the University of Minnesota visited Keech at her home, though without revealing their professional identities or the nature of their interest in the newspaper report that brought them there. As the researchers noted later in their formal account of the episode, they had been studying the behavior of members of millenarian movements whose prophecies failed to materialize but had previously been able to examine such behavior only in the historical record. Alerted to

Keech's prophetic message, however, and to the existence of a group of people following her, the psychologists saw it as an opportunity to conduct a "field test" of their theories under what they described as "natural conditions."[2]

As it developed, the test would have some distinctly unnatural or at least irregular elements. Indeed, it was something of a methodological mare's nest and, as we would see it today, ethical morass, with members of the Minnesota team and as many as five hired participant-observers joining the group around Keech and pretending to be, along with her, convinced to one degree or another of the coming deluge and also, as other messages promised, of their own deliverance to a safe place elsewhere. Among several questions raised but not pursued in the formal report of the study was whether the presence of these relatively numerous participant-observers may have affected either the strength of the group's convictions or, crucially, their subsequent persistence.[3]

To be sure, the flood and rapture, awaited on a hilltop that cold December morning, did not occur. And, just as the team of psychologists had predicted and expected, Keech and several of her followers found ways to explain the nonevent that permitted them to remain convinced of both the authenticity of the messages and the eventual realization of the prophecy. Keech herself seems to have been persuaded that the group's faith in the communications had itself postponed the flood and she reported receiving a specific message to that effect from the extraterrestrial beings they called "the Guardians."[4] No less significantly, however, the phenomenon of scientific interest here—that is, the persistence of belief in the face of apparent empirical disconfirmation—was exhibited by the scientists investigating that phenomenon as well as by their experimental subjects. For, like the expected flood and rapture, a number of the scientists' predictions also failed to materialize. Thus, although—as predicted—Keech's group did not simply shrug off their prior convictions or disband *en masse* after their disappointment, nevertheless, and significantly enough for the theory being tested, several members did loosen their commitment immediately and several others drifted away soon afterward. Moreover, contrary to a key prediction of the theory, only a few of the remaining members of the group attempted to proselytize new members, and the circumstances of those attempts made their status somewhat dubious.[5] Like their subjects, however, the scientists exhibited considerable resourcefulness in explaining the relevant disparities of expectation and experience—in the scientists' case, largely by discounting recalcitrant portions of the data as "unclear" and, in the formal conclusion of their report, describing and interpreting events in ways that minimized their difference from what had been predicted.[6]

We may draw a number of lessons from this story but must bear in mind its sequel. The chronicle of "the Searchers," as Keech and her followers called themselves, ended in 1956, when the team of psychologists studying them published an account of their research in a book titled *When Prophecy Fails.*[7] Although, as just indicated, a number of the empirical predictions of the general theory being tested were not confirmed, the theory itself survived and flourished. Indeed, as elaborated a year later by the team's leader Leon Festinger under the label "cognitive dissonance," it remains one of the most firmly established, highly respected, and intellectually fertile theories in the history of social psychology. Numerous subsequent experiments have confirmed the existence of the central behavioral phenomenon it identifies, that is, people's tendency to persist in their convictions in the face of apparently contrary evidence.[8] No less significantly, related observations and experiments in social psychology and other fields have corroborated the major claim of Festinger's theory as such, that is, that the psychic discomfort we experience when our perceptions clash with our prior beliefs makes the avoidance or attempted reduction of such dissonance a central motive in human behavior.[9]

COGNITIVE DISSONANCE, COGNITIVE CONSERVATISM

The morals to be drawn from this story are, then, both numerous and (as is often the case with parables) somewhat ambiguous. First, the tendency to retain one's beliefs in the face of what strikes other people as clear disconfirmation appears to be a very general phenomenon. It is not restricted to the naive or uneducated, nor is it altogether eradicated by scientific training or good-faith efforts at objectivity. As indicated by related experiments (not to mention the historical record or anyone's casual observation), once we have framed an explanation of some puzzling phenomenon, we are strongly inclined to be most alert to what confirms it. The physician will tend to find alternative diagnoses of a set of symptoms implausible; the scholar will find different interpretations of a text strained. Comparably with regard to broader systems of belief: we generally experience "solid contrary arguments" or "manifestly conflicting evidence" not as potential illuminations but as cognitive irritants. We respond to them not by gratefully adjusting our views (to accord better, it might be said, with "reason" or "reality") but by seeking to remove the irritants. Thus we might discredit the argument as "fashionable," "ideological," or "self-interested" and downplay the significance of the evidence as "dubious," "manipulated," or "a fluke." We may also retain our beliefs under such conditions by more or less subtly altering the form in which we frame them so as to diminish the appearance—to ourselves as well as to others—of outright collision or conflict. These strategic responses to cognitive dissonance are illustrated readily in the apologetics of theists and theologians confronted by the arguments of skeptics or the contrary findings of scholars and scientists. But such self-salvaging responses are also found in the work of skeptics, scholars, and scientists themselves.

Second, although the tendency to belief-persistence is often clearly disadvantageous to the believer, it cannot be considered simply an intellectual defect or pathology as suggested by such terms as "blindness," "bullheadedness," or "irrationality." Rather, it must be seen as fundamentally ambivalent in its operations. Sometimes, in some ways, the tendency—I refer to it as "cognitive conservatism"—is personally limiting or injurious, immuring us in self-sustaining dogmas and risking our crucial unresponsiveness to novel and changing conditions. At other times, in other ways, it is advantageous, ensuring the stability of our empirical generalizations and preventing the possibly costly volatility of our ideas, impressions, and related behaviors. As the Minnesota story indicates, the efforts we expend to conserve our established beliefs may have considerable benefits, practical and intellectual as well as psychological or emotional. It was just as well that Festinger and his team did not toss out their theory of cognitive dissonance just because certain of its specific predictions failed to materialize in a particular test. Indeed, as noted by Thomas Kuhn and other observers in regard to the often cavalier treatment of empirical anomalies by researchers working in an established paradigm, this conviction-preserving streak among scientists seems to be crucial for the general enterprise of science, sustaining the self-confidence, optimism, and energy that also lead, on occasion, to paradigm-changing revolutions of thought.[10]

In the story of prophecies and predictions just told, the positive value of cognitive conservatism is illustrated most clearly by the subsequently vindicated convictions of the Minnesota psychologists and its negative value by the costly credulity of the Searchers.[11] But that contrast is not my point. There are, of course, important differences between the historical trajectories of the empirical sciences and those of popular belief systems, but those differences of ongoing development and outcome cannot be attributed to the distinctive cognitive processes of scientists as compared with those of nonscientists. The relation between scientific achievement and human cognition is a question to which we shall return. For the moment, the point to be noted is that the benefits of belief-persistence—long-range as well as immediate, communal as well as personal—are certainly not confined to science.

A number of cognitive tendencies closely related to belief-persistence will be of continuing interest to us here. Among these are "selective perception," that is, our tendency to notice most readily what specifically concerns our selves and our projects, and "confirmation bias," our tendency to register and remember experiences that confirm our convictions and, correspondingly, to overlook and forget those at odds with them. Other such tendencies include our likely miscalculation of particular forms of probability.[12] Seeking to explain the existence of such problematic dispositions, some cognitive theorists suggest that, although troublesome under many conditions, they may have general advantages for the individual. Thus it is suggested that our intuitive way of estimating probabilities may have permitted our ancestors, and may still permit us, to make the most rapid—and, often enough, close enough—estimates of the odds.[13]

Similar observations may be made regarding cognitive conservatism and also help us understand its ambivalent operations. While cognitive plasticity is obviously necessary for any creature that survives, as human beings do, by learning, the counter-tendency, that is, mechanisms that foster the stability or tenacity of acquired beliefs, would also be necessary and, under many conditions, advantageous.[14] Thus it may be supposed that humans are by nature both docile and stubborn. Much of intellectual life and many episodes in intellectual history, including the history of both science and religion, can be understood as products of the interplay of these reciprocal cognitive propensities *and* of the divergent perspectives from which their operations may be assessed. Given our unusually well developed capacity to learn, we can be duly informed and (it may appear) "enlightened"—but, equally, to be misinformed and (it may appear) "corrupted" or "indoctrinated." Given our unusually well developed inclination to hold fast to what we believe, we can be (what is viewed as) courageous in defense of "truth"—but, equally, to be (what is viewed as) stubborn in attachment to "error" or "heresy."

Much of the formal machinery of logic and many disciplinary norms of science, philosophy, and humanistic scholarship—for example, controlled experimentation, scrupulous self-effacement, or conscientious citation of sources—can be seen as directed toward minimizing the unwelcome effects of these ambivalently operating cognitive tendencies. Indeed, the essential pathos of intellectual history could be seen as our continual effort to clear our cognitive lenses and our continual discovery that we have not managed to do so as well as we had hoped or claimed. There are good reasons, of course, why we pursue such efforts, why we try to exercise maximal rigor and to manifest maximal rationality. But the foregoing observations suggest that there are also good reasons why such hypertrophic efforts always do, in fact, fail and, moreover, carry in their train some unhappy effects. For, given the fundamental ambivalence of the operations of the cognitive mechanisms in question, we cannot eliminate their negative effects, that is, the possibility of indoctrination or persistence in error, without risking the loss of their positive and indeed vital effects along the way, that is, the possibility of enlightenment and the exhibition of courage in defense of what we see as truth.

A comparable set of observations may be made regarding religion, and for comparable reasons. As we shall see in the following two chapters, contemporary research and theory suggest that the cognitive tendencies that give rise to much of what we call religious behavior, from the positing of superior invisible beings to the performance of ritual sacrifices, are indistinguishable from the capacities and dispositions that give rise to what we call culture more generally. Thus it appears that we could not eliminate the conditions responsible for religion and, with it, the recurrent emergence of some of its most troublesome features without risking the loss of much that we value in culture and, with it, the conditions for human existence.

Another point may accordingly be added. Given the fundamentally ambivalent operations of cognitive conservatism as described here, it is not surprising that lists of the individual costs of religious commitment and of the many crimes against humanity committed under its sway can always be countered with equally long and impressive lists of the personal benefits of religious faith and of the many achievements for humanity performed by those inspired by it. Nor is it surprising, given the demonstrated general power of cognitive conservatism, that the final tallies in such cost-benefit assessments appear to have everything to do with the prior cognitive commitments of those doing the tallying.[15] The point will concern us in later chapters in connection with current efforts to explain the persistence of religion in cognitive-evolutionary terms and to defend or question the related idea (and prediction) that the gods are "here to stay."

PERSPECTIVES ON COGNITION

Contemporary cognitive science, along with such more or less traditional fields as epistemology and philosophy of science, is concerned with how we acquire knowledge of the world—or, alternatively phrased, how we construct our worlds and come to operate in them more or less effectively. As the alternative phrasing here indicates, a number of significantly different approaches to human cognition are currently available; and, indeed, several other fields, including some fairly recent ones, such as artificial intelligence, neuroscience, and the sociology of knowledge, are also involved in its study. It will be useful therefore to indicate the particular approach to cognition represented by this book.

Here, as in previous works, I view cognitive processes in

what is sometimes called an "ecological" or "dynamic" framework. That is, like a number of theorists in fields ranging from developmental psychology to the philosophy of biology, I understand human cognition as the full range of processes and activities through which, as embodied creatures, we, like other organisms, interact more or less effectively with our continuously changing environments, thereby ourselves changing more or less continuously. Neither our individual organic structures nor how they will develop are fixed at birth. On the contrary, throughout our lives we interact with our environments in ways that continuously modify our structures and how they operate, and these structural and functional modifications affect our subsequent interactions with our environments, both in what we perceive and in how we behave. It is not merely that our structures define what we can "detect" about "the world" (as in comparisons between the sensory or cognitive apparatus of humans-in-general and bats-in-general) but that the world that each of us (individual human or individual bat) occupies—what we can *act on* and be acted on *by*—is a particular, more or less highly individuated, perceptual-behavioral niche. The continuous mutual process of environmental interaction and organic-structural modification described here is what we commonly refer to as "learning" or "development" and sometimes as "cognition." The structural and perceptual-behavioral modifications themselves, when they are relatively stable and available to self-observation and verbal articulation (as sometimes in humans), are what we commonly speak of as (acquired) "knowledge" or as (changed) "beliefs."[16]

In accord with the framework indicated above, I do not identify cognition or cognitive processes with *mental* as distinct from bodily activities. Nor do I see those processes located in some especially *interior* space, whether the mind, brain, or nervous system, as distinct from the total embodied organism. Nor do I conceive the relevant activities or processes, whether intellectual, psychological, physiological, or neural, as essentially *computations*, that is, as discrete rule-governed operations performed on items of presumptively objective, autonomous information or "input." Also, although I see various cognitive-behavioral tendencies as emerging more or less reliably in humans across history and in all cultures, I do not see them as *innate* in the sense of uniformly present at birth and/or genetically specified in all humans. Finally, although I presume that the neural and other physiological structures involved in the emergence of such cognitive-behavioral tendencies have evolutionary histories, I do not see either of them—the evolved structures or the emergent tendencies—as *fixed* largely in the forms they had in the brains, bodies, or behavior of our stone-age ancestors coping with stone-age conditions. I shall elaborate some relevant implications of this ecological-dynamic view of human cognition as we go along. The important point to be registered here is that, in a number of significant respects, it differs from the strongly mentalist, computationalist, and nativist views that currently dominate the field of cognitive science and that figure accordingly in what are described as "cognitive" approaches to religion.

The view of cognition just outlined relates primarily to its *microdynamics:* that is, to cognitive processes occurring in individual human beings over the course of their lifetimes. But human cognition is also manifested at other, broader levels, where processes of knowledge-formation involve social collectives and occur over historical time. At these levels, or with regard to the *macrodynamics* of cognition, the processes involved are largely the concern of fields such as anthropology, intellectual history, and the philosophy, history, and sociology of science. Here, in company with many theorists working in the fields just mentioned, I regard human cognition from what could be called a constructivist-pragmatist perspective.[17]

One of the most comprehensive and influential articulations of such a perspective is Ludwik Fleck's classic study *Genesis and Development of a Scientific Fact*, originally published in 1935. Fleck, a practicing biologist (his field was immunology) and scholar of medical history, sought to describe the formation and stabilization of scientific knowledge in relation to the dynamics of human cognition more generally. Of particular interest here is his identification and analysis of what he called the "tendency to inertia" of belief systems. "Once a structurally complete and closed system of beliefs consisting of many details and relations has been formed," he writes, "it offers tenacious resistance to anything that contradicts it."[18] Fleck goes on to relate this inertial tendency to what he called the "harmony of illusions": that is, the ongoing mutual attunement of beliefs, perceptions, and material practices that, in his view, characterizes the fundamental process by which epistemic communities—or, in his term, "thought collectives"—maintain their characteristic modes of interpretation and explanation or "thought styles." What follows from this understanding of belief systems for Fleck is a point central to contemporary constructivist epistemology and related historical and sociological views of scientific knowledge. It is that what we come to call the truth or validity of some statement—historical report, scientific explanation, cosmological theory, and so forth—is best seen not as its objective correspondence to an autonomously determinate external state of affairs but, rather, as our experience of its consonance with a system composed of already accepted ideas, already interpreted and classified observations, and, no less significantly, the embodied perceptual and behavioral dispositions that are thereby engendered and constrained.[19]

THE DYNAMICS OF BELIEF

The perspectives on the processes of human cognition outlined above have significant implications for current controversies over the relations—actual and proper—between science and religion, especially insofar as they involve the concept of belief. Most of what we speak of as "beliefs" (commonsense notions, personal assumptions, political, philosophical, and religious convictions, and so forth) can be seen to operate not as discrete proposition-like statements about the world but, rather, as more or less continuously shifting—strengthening, weakening, and reconfiguring—elements of larger systems of linked perceptual-behavioral dispositions. Several features of these systems or, as they are often called, "networks" are of particular interest here.

One is the high degree of interconnectedness and reciprocal determination of their elements: for example, between, on the one hand, our taken-for-granted assumptions, however acquired and whether or not verbally formulated, and, on the other hand, our tendency to observe, interpret, and respond to phenomena in one way rather than another. Thus whether we see—perceive, interpret, and classify—an array of lights in the sky as a constellation of distant suns, a conclave of minor deities, or a squadron of flying saucers, or, indeed, whether we "see"—in the sense of register—it at all, will depend on, among other things, the nature, strength, and configuration of such prior assumptions. And, reciprocally, our perceiving, interpreting, and classifying a string of lights *as* one or another of these will reinforce those very assumptions and tendencies.

A second key feature of belief systems is their *social* constitution and maintenance: that is, the fact that they are formed and stabilized through our ongoing interactions with, among other things, other people, especially members of our particular epistemic communities. One of the major unsupported predictions of the Minnesota psychologists was that members of the quasi-millenarian group they were studying would, after the failure of the prophesied flood, seek to proselytize new members. The prediction reflected Festinger's idea that the distress of cognitive dissonance is reduced by the social corroboration of one's threatened convictions. The general idea was certainly plausible but Festinger's model of the social operations of cognition was relatively crude. For what appears to restabilize our convictions under such conditions is not simply their affirmation by a certain number of other people but the continuous renewal of the processes of social interaction—physically embodied and pragmatically consequential as well as observational and verbal—through which those convictions were shaped and stabilized in the first place.[20] It may be suspected that one reason why the Searchers failed to sustain the degree of belief-persistence predicted by Festinger and his team was that its members did not constitute a functional epistemic community: that is, the sort of structured social collective out of which an effectively normative system of beliefs could emerge. Rather, given their relatively small number (around two dozen, not counting the researchers in disguise), their informal mode of association, and the brief time during which they interacted (for many of them, only a few months prior to the failed prophecy), they formed what could be called a nonce-group: that is, not even or not yet what might have been labeled a "cult" had their active association and mutually exchanged and acted-upon convictions lasted longer. In this respect, the Searchers were quite unlike the historical apocalyptic or messianic movements on the basis of which the Minnesota psychologists made their predictions.[21] They were even further from the types of established collectives whose communal beliefs and practices were theorized by the classic sociologists of religion, Max Weber and Émile Durkheim—and, it may be added, from the types of established collectives of scientists comparably theorized by the proto-sociologist of science, Ludwik Fleck.

The approach to the dynamics of belief outlined above also permits us to appreciate an aspect of cognitive conservatism that might otherwise appear paradoxical, namely, its *creativity*. For the psychological mechanisms involved here operate not simply or primarily by maintaining our established belief systems but also and perhaps more significantly by incorporating into them whatever comes along: novel ideas, anomalous observations, new practical techniques, and so forth. As Fleck's account suggests, this occurs through a process of mutual accommodation in which established beliefs are reinterpreted even as new percepts are classified and old material practices are modified, so that the "harmony" of the entire system—that is, the experienced consonance among its elements—is maintained. As indicated by intellectual history, the reinterpretations can be substantial and, though motivated by a conservative mechanism, yield highly innovative ideas and practices. Thus religions and other belief systems, in spite of what may be efforts at strict heresy-control or claims of an unbroken tradition, are more or less radically transformed over time. We recall Kafka's parable: "Leopards break into the temple and drink the sacrificial chalices dry; this occurs repeatedly, again and again; finally it can be reckoned upon beforehand and becomes part of the ceremony."[22]

Ongoing examples of such responsive self-transformation can be found in theological reactions to Darwinism. Many nineteenth-century theologians rejected evolutionary theory on the ground that it was absurd to believe that life-forms could emerge from pure chance, without God's hand. Thomas Huxley, replying to such arguments, insisted that evolution, properly understood, was not a matter of pure chance—whereupon the theologians, revising their grounds, argued that, if not chance, then necessity and thus all the more evidence of God's hand. Summarizing this episode and comparable earlier ones, historian John Hedley Brooke remarks that throughout the nineteenth century, "design arguments were remarkably resilient, diversifying to meet the challenge presented by . . . new scientific perspectives."[23] More generally we may note that systemic convictions of all kinds—political, philosophical, and scientific as well as theological—typically diversify to meet the challenge of new perspectives and that they often seize the opportunity thus presented for new forms of self-elaboration as well.

In referring here to our experiences of cognitive consonance and the "harmony" of belief systems, I have sought to stress that these are, precisely, matters of subjective experience. That is, they are not what might be seen as objectively well-ordered features of cognitive states, either neural coordinations or proper logical relations among mental propositions. Indeed, empirically and logically speaking, our individual beliefs are always more or less *incoherent*. We might note here a rather striking feature of the collection of ideas that Marion Keech and her followers in some sense "believed," which included visits to earth by superior beings from another planet, communication by automatic writing, and the imminence of both an apocalyptic geological catastrophe and their own escape and salvation through faith. What is especially notable about the collection, in which a number of popular and more or less secular preoccupations of the era (flying saucers and automatic writing—we are in midwest America in the early 1950s here) are combined with an array of vaguely religious and specifically Christian themes (unearthly guardians, prophecy, apocalypse, faith, redemption, and so forth), is not its naivety but its heterogeneity. That sort of heterogeneity, though perhaps especially vivid in this instance, is nevertheless not unusual. Comparable collections of notions, characterized by fragmentary, inconsistent, and relatively undeveloped ingredients with widely varying provenances and significantly different modes, registers, and degrees of conviction, have been documented by numerous researchers. There is reason, moreover, to think that such deeply heterogeneous and technically incoherent assemblages of ideas and schemas are not confined to the uneducated or unreflective but comprise the operative understandings of the world, nominally religious or otherwise, by which each of us lives.[24]

This last aspect of human cognition—the evidently inevitable incoherence of our beliefs—is notable here for two reasons. First, it is rarely considered by those who pursue and promote so-called cognitive approaches to religion. Insofar as those approaches participate in dubious rationalist-logicist views of the nature of belief, they remain severely limited in their observations and accounts of the psychological operations of religion.[25] Second and conversely, due recognition of the complex structure of human belief affords an important perspective on the operation of formal creeds in the lives of religious people and also on such chronic issues as the compatibility of personal religiosity with scientific knowledge or the more general survival of religion in a secular world.

PROPHECIES AND PREDICTIONS

To the extent that Western science operates as a comprehensive belief system, one of its central features is the confidence it encourages in its eventual triumph over all other such systems. Nowhere has that confidence been expressed more forthrightly than in the so-called secularization thesis—that is, the idea, advanced by many twentieth-century social theorists, that religious beliefs and institutions tend to weaken with modernization and, with the global spread of literacy and technology, will sooner or later disappear. Few general theories in the social sciences have been more self-assuredly maintained than this one, and few world-historical predictions have failed more spectacularly, at least so far. As late as the 1960s and in the face of what is now recognized as an upsurge of new religious or quasi-religious movements in the United States, including UFO cults and New Age spirituality, social scientists were still describing the correlation of modernization and secularization in pretty much the same terms as Weber or Freud more than half a century earlier. Here is anthropologist Anthony Wallace making predictions in 1966: "Belief in supernatural beings and in supernatural forces that affect nature without obeying nature's laws will erode and become only an interesting historical memory. To be sure, this event is not likely to occur in the next hundred years. . . . But as a cultural trait, belief in supernatural powers is doomed to die out, all over the world, as a result of the increasing adequacy and diffusion of scientific knowledge."[26] Even among those social scientists who gave some thought to religious developments in the non-Western world (a group that included Wallace himself), none foresaw either the emergence of so-called radical Islam and Hindu nationalism or the rapid spread of Pentecostalism in Africa and Latin America.

Current efforts to explain the persistence of religion will concern us in later chapters but, in connection with that signal failure of foresight, a comment by sociologist James Beckford on reactions to it by his fellow social scientists is especially apt here. He writes:

> The debates about secularisation reveal very few cases of willingness to abandon fundamental points of view or . . . to take account of the known arguments. Instead, it is more common for social scientists to defend their particular concepts and interpretations, in the face of arguments that are allegedly fatal to them, by re-specifying the meaning with which they invest their concepts. . . . The possibility that fresh evidence might actually persuade social scientists explicitly to abandon a concept such as secularisation is almost inconceivable.[27]

This recalls a key passage in Fleck's *Genesis and Development of a Scientific Fact* on the inertia of belief systems: "When a conception permeates a thought collective strongly enough, so that it penetrates as far as everyday life and idiom and has become a viewpoint in the literal sense of the word [that is, a way of perceiving], any contradiction appears unthinkable and unimaginable."[28] For Beckford as for Fleck, whose understandings of the general dynamics of human cognition are quite similar, this intellectual intransigence is altogether to be expected. Thus Beckford concludes: "There is nothing abnormal or shocking about this. . . . High-level abstractions [such as "secularization" or "modernization"] operate like compressed narratives and, as such, can be re-written or re-interpreted to accommodate facts or ideas that might appear to be inconvenient. This does not mean that the narratives are entirely fictional or mythical. . . . But the relationship between adherence to the narrative and openness to the 'evidence' . . . is extremely elastic."[29] Current social-scientific accounts of religion, now characteristically seeking to explain "why gods persist" rather than predicting their disappearance, can be regarded as duly chastened efforts.[30] Nevertheless, what Beckford calls "adherence to the narrative," here the paradigmatic cognitive-evolutionary scenarios that unite and define these New Naturalist accounts, continues, as we shall see, to bear a rather elastic relationship to the evidence.

Beckford's reference to the rewriting of narratives "to accommodate facts or ideas that might appear to be inconvenient" returns us to what I described above as the paradoxical-seeming creativity of cognitive conservatism, here in regard to religious writing. In a book published twenty-five years after *When Prophecy Fails*, this one titled *When Prophecy Failed*, biblical scholar Robert Carroll invokes Festinger's theory of cognitive dissonance to elucidate the composition of the prophetic books of the Hebrew Scriptures. Carroll's central point, elaborated through a close reading of the text of Isaiah, is that a major creative force in the prophetic tradition, generating cascades of new textual interpretation and reinterpretation, was the community's experience of cognitive dissonance when its oracles failed to materialize. As he puts it in a suggestive formulation, "*dissonance gives rise to hermeneutic.*"[31] More generally, as widely noted and variously explained, when convictions based on a sacred text are contradicted by empirical findings, the convictions are rarely rejected outright. But, exemplifying the cognitive dynamics outlined in this chapter, the convictions in question are not just maintained unchanged either. Rather, as the empirical findings are interpreted and incorporated, the sacred text is reinterpreted, elaborated, and modified so that the systems of belief in relation to which both—the findings and the text—are perceived continue to be experienced as harmonious, or, it could be said, so that the book of Nature and the Book of God continue to tell the same story. As we shall see, that is precisely what a number of contemporary theologians seek to demonstrate in what I call here the New Natural Theology.

TU QUOQUE

A recurrent ploy in theistic rejoinders to polemical atheism and in theological replies to naturalistic accounts of religion is a charge of *tu quoque:* "You, too! You yourself do just what you charge us with doing." Thus the local priest may accuse the local atheist—just as the atheist accuses religionists—of mindless conformity to received ideas. Or the theologian may note that the naturalist's naturalism, just like the theologian's theism, can be accounted for naturalistically. In form and circumstance, the *tu quoque* charge seems to be both universal and primitive. The pot is reproached everywhere for calling the kettle black. To the perennial schoolyard taunt, "Your mother is . . . ," the perennial schoolyard retort seems to be, "So's yours!" On theological occasions, the charge is often produced as a *coup de grace*, exposing a mirror-like duplication at the heart of the atheist-naturalist's insulting or reductive characterization of religionists and thus an implicit self-accusation and, with it (or so it is maintained), a conclusive self-refutation.[32]

There is, of course, a similarity between the theologian's charge of *tu quoque* and my observations above and elsewhere in this book regarding the duplication by social scientists of the behavior and presumed cognitive processes of their credulous subjects. But there are several crucial differences between the charge and my observations here. First, I am not concerned to indicate a correspondence of putative disgraces, as between the blackness of pots and of kettles. As emphasized above, the shared cognitive processes involved here are not pathologies; they operate ambivalently—for both good and ill—in all domains. Second, contrary to the implication that often attends the theological charge, my aim here is neither to demote the credibility of science nor, thereby or otherwise, to boost the credit of any theistic or theological claim. To observe that scientists display some of the same cognitive tendencies that evidently give rise to and sustain religious beliefs is not to dispute the operational validity of scientific knowledge. As we learn with increasing force from the history of science and studies in the sociology of scientific knowledge, the authority of the scientific enterprise does not rest on the deployment of other-than-human cognitive capacities or require the presumptive transcendence of ordinary human limits.[33] Finally, in noting that the practices and promotions of science involve cognitive and logical-rhetorical operations also found in religion and theology, I do not seek to show that science is, as is said, "just another religion." What I do seek to suggest, rather, is that science, religion, and human cognition are each more copious than commonly recognized and more complex than sometimes theorized.

CHAPTER TWO

Cognitive Machinery and Explanatory Ambitions: The New Naturalism

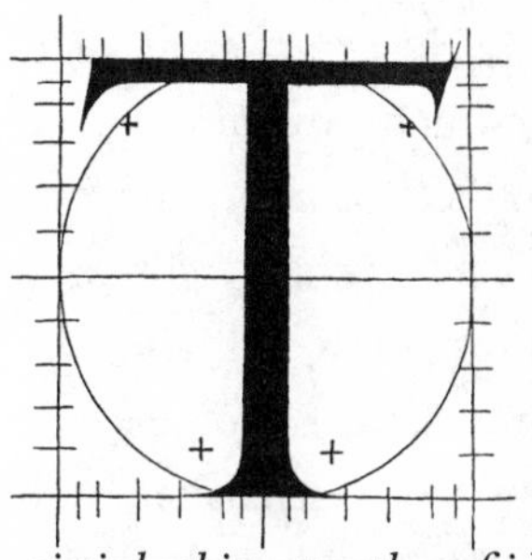

These English psychologists—what do they really want? One always discovers them . . . at the same task, . . . seeking the truly effective and directing agent . . . in just that place where the intellectual pride of man would least desire *to find it . . . for example, . . . in a blind and chance mechanistic hooking-together of ideas, or in something purely passive, automatic, reflexive, molecular, and thoroughly stupid. . . . Is it a secret, malicious, vulgar, perhaps self-deceiving instinct for belittling man?*

But . . . if one may be allowed to hope where one does not know, then I hope . . . that these investigators and microscopists of the soul may be fundamentally brave, proud, and magnanimous animals, who . . . have trained themselves to sacrifice all desirability to truth, every *truth, even plain, harsh, ugly, repellent, unchristian, immoral truth.—For such truths do exist.*

—*Nietzsche,* On the Genealogy of Morals

Efforts to explain religion, as it is said, "scientifically" are of considerable interest beyond the fields most immediately involved—that is, the academic discipline of religious studies and, currently, such other fields as anthropology, cognitive science, and evolutionary psychology. For such efforts, by their very existence, engage a number of issues that have become central to contemporary thinking about intellectual life. These include the relations, actual and proper, between the natural sciences and other disciplines and the role of science, limited or expansive, in the study of human phenomena. In connection with such issues, I review here some familiar perplexities presented by the longstanding effort to explain religion naturalistically and go on, in this chapter and the next, to consider a set of recent works representing self-described "cognitive" and "evolutionary" approaches to that project. In assessing these and related works, which explain religious concepts and practices largely as products or byproducts of the operation of evolved cognitive mechanisms, I compare them with other naturalistic studies of religion, past and current. As I seek to suggest in both chapters, the ongoing effort to explain religion naturalistically is worth our attention, both as a significant project in Western intellectual history and as a provocative problem for epistemology, the philosophy of science, and the sociology of knowledge.

NATURALISM AND THE STUDY OF RELIGION

There is no reason to doubt, though it has been doubted, that accounts of religion can be given in strictly naturalistic terms—that is, without appeals to entities or forces outside of what we think of as nature.[1] Such accounts, dealing both with exotic beliefs and practices and with religious forms closer to home, have been offered since antiquity. The most significant of them make up an intellectually impressive roll call: Lucretius's *De Rerum Natura*, David Hume's *The Natural History of Religion*, Nietzsche's *On the Genealogy of Morals*, Max Weber's *The Sociology of Religion*, Émile Durkheim's *The Elementary Forms of the Religious Life*. Intellectually impressive as all these are, however, their adequacy, along with that of other naturalistic accounts of religion, can be—and has been—questioned from a number of angles.

First and most obviously, the phenomena that make up what is called, in most of these works, "religion" are exceptionally heterogeneous, ranging from individual experiences and popular beliefs to formal doctrines, ritual practices, social institutions, and political effects. Moreover, they include, at least presumptively, practices from ancient times to the present, in all types of cultures and societies and across all regions of the globe—from Neanderthal burial rites to Vatican encyclicals. Indeed, it is not clear that the term "religion" names a coherent set of phenomena at all or whether, like "art," "culture," or "literature," it is not a shifting artifact of the various disciplines and discourses that take it as their nominal subject.[2]

The question of coherence inevitably haunts the field of religious studies and contributes, it seems, to an ongoing sense of crisis among its practitioners, though not to the point (at least not yet) of dissolving the discipline itself. In any case, in the study of religion as in any effort to explain a manifestly vast and complex array of phenomena, different levels of analysis and theoretical perspectives can always be brought into play. Moreover, and especially significantly here, different broader purposes can also be invoked, each implying different criteria of assessment and, with each, different rhetorics or styles of presentation. Accordingly (or so it would seem), no "explanation of religion," naturalistic or other, could properly claim to be complete, final, most fundamental, or most genuinely explanatory. But, of course, that does not keep theorists from claiming just such status for their particular accounts.

In addition to limits of the kind just described, common to intellectual projects of any scope, naturalistic accounts of religion encounter difficulties of another kind, these familiar from efforts to explain other highly valued experiences or cultural achievements—for example, consciousness, love, literature, or, in fact (as will be significant later on), science itself—in strictly neutral, objective, physicalist terms. In these other cases, too, rigorously naturalistic descriptions or accounts (for example, explanations of consciousness in terms of patterns of neural firings, monetary cost-benefit analyses of parental sacrifice, or diagrams of social networks among high-energy physicists) are regarded by many people as manifestly inadequate to the experiences, motives, and achievements involved: reductive, ignorant, insensitive, and, commonly, as missing the main thing, whatever that thing is. In the case of religion, it is sometimes referred to as "the sacred."

Views and sentiments along the latter lines were given influential expression about fifty years ago by the Romanian scholar of religion Mircea Eliade, who wrote: "A religious phenomenon will only be recognized as such if it is grasped at its own level, that is to say, if it is studied *as* something religious. To try to grasp the essence of such a phenomenon by means of physiology, psychology, sociology, economics, linguistics, art or any other study is false; it misses the one unique and irreducible element in it—the element of the sacred."[3] Many other scholars in the field have concurred, at least to the extent of regarding religious studies as clearly a humanistic discipline, not a natural science or, for that matter, even properly a social science. The question of appropriate method, however, is by no means settled and continues to generate disputes among practitioners. The disputes are often exacerbated, it seems, by personal ambivalence. Thus a practitioner observes that scholars in his field typically *like* religion but *respect* science and are, accordingly, faced with a set of emotional-intellectual conflicts that translate into methodological dilemmas. As he sees it, the difficulty comes down to a choice, always experienced as to some extent unsatisfactory, between, on the one hand, offering a duly sympathetic but not entirely objective "appreciation" of religion and, on the other hand, producing a duly objective but alienating scientific—or quasi-scientific—"analysis" of it.[4]

While these conflicts and related disputes between advocates of one or another method in religious studies ("phenomenological" versus "scientific," "empathic" versus "objective," "interpretive" versus "explanatory," and so forth) have been framed and negotiated in various ways over the past half century, they have recently become considerably sharper. Thus a number of science-identified scholars of religion, appealing to the achievements of nineteenth-century founders and charging their colleagues with failings ranging from anti-intellectualism to crypto-theology, have sought, as one of them puts it, "[to recover] the scientific agenda in the academic study of religion, on which basis it achieved cognitive legitimation."[5] Responses to such charges and programs run the gamut from mild concession and genial self-defense to strong resistance and scorn. In an especially sharp but also especially interesting response, theologian Paul J. Griffiths, commenting on the call for a scientific agenda just quoted, maintains that "[any] attempt to make sense and use of an idea of religion that systematically rejects theological assumptions will fail."[6] Citing the difficulties presented to such efforts by "the very idea of religion"—an idea, Griffiths argues, that "is so deeply intertwined with" something "theological and specifically Christian . . . that attempts to disentangle it by abstraction will produce a cipher"—he concludes: "[T]hat this is so shows that the scientific study of religion is without a future."[7]

Griffiths may be right about the intellectual strains involved in any generalized account of what is termed "religion." (Contemporary theorists of the subject often make the point, though commonly from a different perspective.)[8] Nevertheless, his prediction appears precipitous. Certainly for researchers working in fields such as anthropology, psychology, or sociology and for whom religion is an object of ongoing intellectual and sometimes programmatically "scientific" interest, such sentiments are irrelevant and such efforts at confinement increasingly objectionable. Indeed, the past fifteen years or so have been a period of extraordinary activity in pursuit of just that programmatic interest, with the appearance of a number of self-consciously innovative and science-identified books with titles such as *How Religion Works: Towards a New Cognitive Science of Religion* or *Why Gods Persist: A Scientific Approach to Religion.*[9]

Some idea of the revolutionary temper of these works can be gathered from the opening words of one of the earliest of them, *Rethinking Religion,* co-authored by E. Thomas Lawson and Robert N. McCauley and published in 1990. "Some books," they write, "make trouble. It is certainly the aim of this one to do so." As Lawson and McCauley go on to explain, the trouble that they seek to make is specifically for the field of religious studies, which they represent as enfeebled by the timidity and torpor of scholars "who deny the possibility of fruitful scientific explanation of wide domains of human action and experience" and "[whose] ambitions extend no further than contorted taxonomies and thick descriptions."[10] The latter two allusions are evidently to, respectively, comparative religion in the tradition of Eliade and cultural anthropology in the tradition of Clifford Geertz. Since, together, these two approaches cover quite a large territory in the contemporary study of religion, clearly something seriously disruptive is being proposed.

A certain amount of fractious self-positioning is a recurrent, though not universal, feature of the works just mentioned. What unites them more fundamentally and significantly is, first, the centrality for their approaches of methods, studies, and theories drawn from evolutionary psychology and the rather sprawling field of "cognitive science" and, second, their more or less strenuous identification of their projects with "science," itself rather monolithically and sometimes triumphalistically conceived. It is in these two respects—that is, an explicit orientation toward a particular type of psychobiological/adaptationist explanation of human behavior along with a tendency toward an earnest and sometimes aggressive scientism—that these works constitute what I am calling here the New Naturalism in the study of religion.[11]

Given the multiple, shifting meanings of the term "scientism," I would note that, in my own usage, it is not equivalent to the valuing of empiricism, systematicity, or logical rigor in the pursuit of knowledge or to a generally appreciative view of the activities and achievements of Western science.[12] Rather it refers to the conviction, articulated influentially by Edward O. Wilson in his 1998 book *Consilience: The Unity of Knowledge*, that the aims, methods, and products of the natural sciences should be taken as models for all knowledge practices—a conviction often attended (as in Wilson's book) with the idea that the humanities disciplines are at best pre-scientific and should be shepherded as quickly as possible, along with some still vagrant social sciences, into the fold of the natural sciences. I have argued against those convictions and programs elsewhere.[13] What I would stress here is that, contrary to the evident assumption and frequent charge of some promoters of New Naturalist approaches to religion, a rejection of scientism in the sense just given need not be theologically motivated or entail a general hostility to the naturalistic study of human phenomena. Conversely, and contrary to the charge and practice of many broadly humanistic (as well as specifically theistic) critics, the fact that a study of some aspect of human culture or behavior, including religion, draws extensively on current scientific findings or even that it is programmatically naturalistic would not, in my view, warrant the label "scientism" or make such a study dismissible on just that count.[14]

Cognitive-evolutionary explanations of religion deserve and require careful review and discriminating assessment. The intellectual interest of the New Naturalist project and its promise for affording better understandings of important features of human behavior and culture should, I think, be recognized. But I also think that critical attention should be given to the intellectual confinements represented by some of the project's characteristic methodological, conceptual, and also, in effect, ideological commitments, especially when these are viewed in relation to alternative current theories of human behavior, culture, and cognition and to the entire naturalistic tradition in the study of religion.

COGNITIVE MACHINERY

One of the most influential works of the New Naturalism is a book by anthropologist and evolutionary psychologist Pascal Boyer titled, with imposing finality, *Religion Explained.*[15] Boyer opens it with a sharp contrast between, on the one hand, explanations of religion based, like his own, on "findings and models in cognitive psychology, anthropology, linguistics and evolutionary biology" and, on the other hand, what he refers to as "intuitions," "spontaneous, commonsense" ideas, and "most accounts of the origins of religion." Though he gives few explicit examples of the last of these, it is evident from his descriptions and incidental references, including an allusion to "bookshelves . . . overflowing with treatises on religion, histories of religion, religious people's accounts of their ideas, and so on," that they include not only works by scholars such as Mircea Eliade or Paul J. Griffiths, who see religion as immune to naturalistic explanation, but also the classic naturalistic accounts mentioned above by, among others, Hume, Weber, and Durkheim. "Spontaneous" and "commonsense" are of course peculiar terms to use to describe accounts that offer, as Hume, Weber, and Durkheim do, considerable erudition, extensive empirical observation, detailed analysis, and more or less unconventional conclusions. For Boyer, however, the crucial contrast is between explanations of religion that are genuinely "scientific"—in a specific sense I discuss below—and everything else. According to that measure, all the items on those overflowing bookshelves are equivalent to each other and also to anyone's spontaneous, commonsense ideas about religion. "All," he declares, "fail to tell us why we have religion and why it is the way it is."[16]

In elaborating the contrast, Boyer produces a typology of failed accounts of religion and, for each, a corrective alternative in terms of some set of evolved, adaptive, mental mechanisms. Thus, against the "intellectualist" view that religions offer people ways to understand significant but puzzling phenomena, he informs readers (the didactic mode is persistent): "Our minds are not general explanation machines. Rather, minds consist of many different, specialized explanatory engines[,] . . . more properly called *inference systems*, each of which is adapted to particular kinds of events and automatically suggests explanations for these events." Similarly, against the "emotive" view that religions allay fundamental human anxieties about death and personal mortality, Boyer writes: "But human emotions are not that simple. They happen because the mind is a bundle of complicated systems working in the mental basement and solving very complex problems." The mental systems do this, he explains, through computational programs that are triggered by specific kinds of situations. After brief discussions of what he identifies as two other common but inadequate explanations of religion, namely, the "social-functionalist" claim that religions hold society together and support morality and the "irrationalist" account which attributes religious concepts to primitive or distorted ways of thinking, Boyer provides a summary of what he will present in the book as the adequate explanation of religious concepts: "Some concepts happen to connect with inference systems in the brain in a way that makes recall and communication very easy. . . . Some concepts happen to connect to our social mind. Some of them are represented in such a way that they soon become plausible and direct behavior. The ones that do *all* this are the religious ones we actually observe in human societies. They are most successful because they combine features relevant to a variety of mental systems."[17]

Boyer's main point in the passages quoted above and throughout *Religion Explained* is that, contrary to more familiar views, we should understand religious ideas (for example, gods, immortality, or moral teachings) and related practices (for example, prayers or communal rituals) not as more-or-less functional (or dysfunctional) human responses to recurrent human conditions and experiences but, rather, as effects of the automatic operation of a number of specific, highly specialized, innate and universal mental mechanisms. In the evolutionary-psychology paradigm to which Boyer subscribes, it is assumed or proposed that these mechanisms evolved by natural selection to provide our ancestors with adaptive solutions to an array of fitness-related problems recurrent under stone-age conditions, for example, foraging, avoiding predators, choosing mates, or distinguishing friend from foe.[18] The mechanisms (or "systems" or "devices") are also posited as being discrete ("modular"), genetically specified, and somehow (though it is not yet known exactly how) neurophysiologically realized—or, in the computer-derived idiom of much mainstream cognitive science, as being "coded" in "our" (presumptively universally shared) genes and "hardwired" in "the" (singular, presumptively universally shared) human brain.

Boyer presents a picture of human behavior as largely a matter of the automatic, unconscious workings of evolved mental mechanisms, and he promotes the description of such workings as a properly scientific explanation of religion that trumps all other accounts. Indeed, for Boyer, it is precisely insofar as an explanation of some phenomenon—any phenomenon—is put in terms of what he refers to repeatedly as "underlying causal mechanisms" that it counts as genuinely scientific. I turn below to how these views operate in his explanations of specific aspects of religion, but, in relation to these central features of *Religion Explained* and of self-described "cognitive" accounts of religion more generally, two important points should be made.

First, it should be recognized that neither the computational-modular model of mind nor the idea of innate, automatically triggered mental mechanisms is a foregone conclusion of contemporary cognitive science or of any other science. The computational model has been significantly challenged both by practitioners of "cognitive science" per se and by researchers and theorists working in related fields, including evolutionary biology, linguistics, developmental psychology, neuroscience, paleoanthropology, and philosophy of mind. Moreover, as indicated in the previous chapter, a number of important alternative models of cognition have been developed in these and other fields.[19] These models often give considerable attention to a number of features of human cognition slighted in Boyer's book and in the new cognitive accounts of religion more generally. Among them are the significance, for humans, of ongoing individual experiential learning; the complex social, physical, and pragmatic dynamics involved in the transmission of skills and beliefs; and the presence among post-Paleolithic humans of such crucial cultural cognitive resources as transgenerational material culture, schools, texts, and duplicated images.[20] Such alternative models also give due attention to the significant differences among individuals with regard to various aspects of cognition. Contrary, then, to the assumptions of paradigmatic evolutionary psychology and the claims of current cognitive explanations of religion, it is by no means clear that our interactions with our environments are determined largely by the operation of mental mechanisms hardwired at birth or that various widespread and recurrent features of human behavior and culture, including those associated with religion, are best explained by reference to a universal and virtually uniform species-specific mind. What Boyer presents in *Religion Explained* as a properly corrective, hard-nosed, comprehensive explanation of religion based on up-to-date, established "evolutionary" and "cognitive" scientific knowledge is better described as an array of more or less speculative accounts of selected features of religious belief and practice based on a set of still highly controversial theories developed in fields at some distance from evolutionary biology and empirical neuroscience.

Second, Boyer's identification of scientific explanation with descriptions of underlying causal mechanisms is questionable both from a theoretical perspective and also in relation to what can be claimed by his own accounts.[21] While the identification reflects standard views formed when the major examples of scientific theories and explanations were drawn from the physical sciences (largely astronomy, physics, and chemistry), current understandings in philosophy of science recognize a variety of explanatory modes and due attention is given to the biological, behavioral, and social sciences.[22] In the latter sciences, explanation often takes the form of models of the emergence of complex phenomena from the dynamic interaction of multiple forces and contingent events operating at various levels of organization. With regard to many aspects of religion, explanations of this kind are likely to be more adequate to the range, complexity, and heterogeneity of the phenomena involved than the strictly linear, input-output, inside-to-outside, depth-to-surface models sought and produced in evolutionary psychology. Indeed, on Boyer's own definitions, his explanations of religion are scientific primarily in aspiration. Thus, while he defines scientific theories as those in which "we describe phenomena that can be observed" and "explain them in terms of other phenomena that are also detectable," in many of his own accounts, as in those of evolutionary psychology more generally, the mental "mechanisms," "systems," and "devices" in terms of which behavioral and cultural phenomena are explained are not observable at all and "detectable" often means little more than hopefully posited and strenuously asserted.[23]

I turn now to some of those explanations.

EXPLAINING GODS

For Boyer, as for New Naturalists generally, the existence of concepts of supernatural beings—for example, gods, spirits, or ancestors—is a defining feature of religion. The soundness of that view, including the presumed transparency and transhistorical stability of the meaning of the term "supernatural," will concern us in later chapters. Here, however, we may consider how such concepts are said to originate. In an account now standard among cognitive theorists of religion, two mental mechanisms are said to be central. One is a system for agent-detection, said to have evolved to alert our ancestors to the possible presence of prey or predators and, even now, hair-triggered by unexpected motions, sounds, or other appearances. The suggestion is that, due to the hypersensitivity of this mechanism, humans are led to posit animate agents of some kind even when none are there, especially when confronted by sudden, striking, and/or otherwise inexplicable events.[24] A second mechanism, which Boyer describes in some detail, is the activation of an "inference system" that automatically supplies such agents with particular characteristics or properties. In the view of human cognition that he follows here, the concepts we can form are limited by a set of basic, hardwired "ontological categories," including *animal*, *plant*, *person*, and *tool*, each of which involves a set of "default properties." Thus, once the activation of our agent-detection mechanism has caused us to posit an agent and on some basis to classify it as *person*, our mind automatically outfits it with such properties as sensory capacities, intentions, moods, and memories. In this way we—or "our minds"—generate watchful ancestors, angry spirits, and jealous gods who, like other agents we classify as persons, we automatically assume may have friendly or hostile intentions toward us and who, as other inference systems kick in from our evolved "social mind," we also automatically assume must be thanked, flattered, feared, and appeased.[25]

As a naturalistic explanation of the origin of ideas of deities, a good bit of this account makes sense and, indeed, has done so for some time, with the tendencies Boyer describes in terms of the automatic activation of discrete mental systems otherwise recognizable under such terms as *animism*, *anthropomorphism*, *personification*, or *projection*, all seen as "natural" and "primitive" features of human psychology. Thus David Hume, in a passage cited by several New Naturalists and sounding here much like one of their company, writes: "There is an universal tendency among mankind to conceive all beings like themselves, and to transfer to every object those qualities with which they are familiarly acquainted and of which they are intimately conscious. We find human faces in the moon, armies in the clouds; and by a natural propensity, if not corrected by experience and reflection, ascribe malice or good-will to every thing that hurts or pleases us. . . . [T]rees, mountains and streams are personified, and the inanimate parts of nature acquire sentiment and passion." Hume goes on to stress the generality of this tendency in an observation omitted by the New Naturalists who quote the preceding passage: "Nay, philosophers [that is, natural philosophers or, as we would say, scientists] cannot entirely exempt themselves from this natural frailty; but have oft ascribed to inanimate matter the horror of a *vacuum*, sympathies, antipathies, and other affections of human nature."[26] A related observation appears in *On the Genealogy of Morals*, where Nietzsche describes our common tendency to posit discrete causes and effects, along with personalized agents ("doers") and purposeful acts ("deeds"), when confronted by phenomena that could otherwise be understood as processes of occurring, unfolding, or becoming. Like Hume, Nietzsche notes pointedly that the psychological tendency in question is also found among scientists, as illustrated, in his example, by their positing of various hidden "forces" said to "move" or "cause" this or that phenomenon.[27] The tendency could also be illustrated by cognitive scientists' division of the ongoing processes of human cognizing and behaving into upper-level "behavioral phenomena" and basement-level "mental mechanisms" and by their positing the unobservable operations of the latter as the hidden causes of the former. Boyer himself notes the generality of this tendency among humans but makes an exception of scientists: "Unless we do science, the way we explain our ordinary intuitions very often refers to inscrutable causal processes."[28]

Hume and Nietzsche were not the only naturalizing theorists of religion to anticipate key claims of the New Naturalism. The observation of a relation between recurrent features of religion and species-wide dispositions was already securely in place by the end of the nineteenth century along with more general speculations on the origins of religion in human psychology. Indeed, Edward Tylor's *Primitive Culture* and James G. Frazer's *The Golden Bough* are, to a large extent, "cognitive" approaches to religion, and Theodor Gomperz, who followed Darwin's work closely, anticipates many quite specific New Naturalist observations and explanations in his account of the origins of Greek religion.[29] I stress the existence of these precursors not to suggest that that there is nothing new under the sun—though, with respect to explanations of religion, that may be broadly true. The emphasis is required, rather, because Boyer and other New Naturalists tend to exaggerate the revolutionary character of their ideas and, in the course of promoting them, produce distorted accounts of the history and current state—and, indeed, existence—of a long naturalistic tradition in the study of religion.[30]

Boyer would insist that there are important differences between earlier naturalistic accounts, which merely describe or "point to" certain "phenomena," and the explanations he offers in *Religion Explained*, which, he claims, identify the underlying mental mechanisms that actually *cause* them.[31]

Significant differences can certainly be observed. In Hume's account, anthropomorphism is described as our general tendency to project onto other beings what we observe and experience as our own features and feelings; in Boyer's account, it is explained as the automatic activation of a set of evolved hardwired inference-systems. Similarly, whereas Nietzsche associated the positing of invisible "doers" and "deeds" with a common tendency to segment and reify processes, Boyer traces it to the hyperactivity of a specific mental module. To the extent that Boyer's accounts are cast in terms of more extensively developed and established theoretical paradigms, they can be said to represent intellectual advances over the insightful but ad hoc observations of Hume and Nietzsche. Given, however, both the fundamental dubiousness of many of the assumptions, models, and methods that Boyer carries over from the evolutionary psychology paradigm and also the speculative character of many of the mental mechanisms he posits and invokes, one would hesitate to say that *Religion Explained* represents a quantum jump from the earlier naturalists' merely "commonsensical" and plainly inadequate "intuitions" about the origin of supernatural concepts to his own certifiably scientific and finally adequate explanation of their causes.

A further point may be added. Many earlier theorists in the naturalist tradition recognized that religions are *dynamic* and represented them as such: that is, as systems of beliefs, practices, and institutions that change over time in response to various historical forces and events. Thus, in the genealogical accounts framed by Hume, Nietzsche, Gomperz, and Weber, concepts of the gods and practices of worship are seen as responding both to significant changes in the social and lived physical environment and also to intellectual, economic, and other cultural developments, including the effects of those religious concepts and practices themselves.[32] These complexly interactive and temporal aspects of religion receive little acknowledgment from Boyer or other New Naturalists, who typically ignore or downplay historical dynamics in favor of what are posited as prehistoric conditions and transhistorical mechanisms. That imbalance of attention will be of continuing concern here.

IMMORTALITY AND MORALITY

Among other topics treated at some length in *Religion Explained* are religious ideas of immortality and the socially normative operations of shared religious beliefs and practices. In an intriguing section, Boyer seeks to demonstrate that concepts of immortality and of a life after death can be explained without reference either to an abstract dread of mortality or to a general human need for consolation at the death of loved ones. Such concepts, he argues, originate rather in a suite of evolved intuitive and emotional responses that are automatically elicited in humans by the presence of a dead body. The responses involved include fear, automatically elicited because, for our ancestors, a corpse was a sign of predators nearby; an avoidance response that evolved to protect us from toxic substances, such as dead bodies; and grief, an emotion triggered by the confusion or frustration of hardwired systems primed for social interaction in the presence of recognized persons. There is, Boyer observes, nothing inherently religious about any of these responses to human corpses. They yield such familiar "religious" notions as undying human souls and a personal afterlife only when they are attached to already existing supernatural concepts, such as immortal gods living in invisible, unverifiable places.[33]

Many aspects of this account are persuasive, among them Boyer's vivid delineation of the intensity, multiplicity, and confusion of our responses to the presence of dead bodies, especially those of persons known to us, and his indication of the relation between those specific responses and more generally recognized human emotions and impulses. As I shall suggest in the discussion below, it is just such features (among others) of elaborated naturalistic accounts of religion—evocations of recognizable human experiences and responses placed in broader intellectual frameworks—that make them, at their best, appear satisfyingly explanatory. Several important points, however, should be added.

To begin with, nothing in Boyer's account negates more familiar views that connect the origin of concepts of an afterlife to people's dread of their own death or pain at the death of others.[34] He simply adds to such views another set of observations and hypotheses at another level of explanation. No less significantly, the elements of Boyer's account to which he gives the greatest emphasis, namely, those most characteristic of a specifically "cognitive" approach, are also the ones that are most empirically and conceptually problematic. These include his suggestion, first, that our reactions to death are *always* and *only* the product of hardwired mental mechanisms and, second, that the mechanisms involved are provoked *directly* by external sensory inputs—in this case, by the presence of a dead body. There is no evidence supporting these overstated suggestions and much evidence supporting the view that our responses to death—including the idea of our own death and our encounters with the bodies of dead persons—are as complexly mediated as most of our responses to the worlds within and around us.

To produce a strictly mental-mechanism account of ideas of immortality, Boyer must ignore both the range of individually differentiated responses to death known to us from our own experiences and from the historical record (travelers' reports, journals, letters, autobiographies, and so forth) and also the extensive cultural elaborations of the concepts at issue—death, immortal spirits, a realm of existence beyond death, and so forth—found in art, literature, philosophy, and social ritual. These aspects of human responses to death and of human responses more generally—that is, their high degree of personal individuation and extensive cultural elaboration—are not merely overlooked by Boyer. They are determinedly excised—or, perhaps, rendered invisible—by the commitments that define his approach, in accord with which deep-seated, innate, presumptively uniform mental mechanisms are given explanatory priority while phenomena that challenge such explanations are treated as insignificant surface variants or mere ancillary effects.

A similarly intriguing but problematic picture of social and so-called moral behavior follows from related theoretical commitments. In current evolutionary and rational-choice models of human behavior, our social interactions are seen as governed largely by intuitive calculations of the advantages we may derive from those interactions individually. To the extent that we behave in moral, cooperative, or altruistic ways at all, it is, according to such a view, because of innate impulses that evolved to promote our own genetic fitness—for example, by promoting the fitness of our gene-sharing kin, or by gaining reciprocated benefits for ourselves, or by helping us secure the advantages of membership in a social coalition. Thus Boyer explains the in-group cooperativeness or sense of mutual connectedness characteristic of human communities as arising not from shared beliefs, shared goals, or participation in communal activities but, rather, from emotions and impulses triggered in individual members by a specific, evolved inference system that he calls our "social mind."[35]

Applying these models of social behavior to religion, Boyer observes that religions and gods serve coalitional purposes, but not in the way commonly thought, that is, as the focus of shared religious beliefs and rituals. Rather, the crucial fact is that we need to signal our coalitional membership clearly to one another in order to know who can be trusted and who cannot. Accordingly, we create highly visible marks of group identity, such as special foods, special languages, or special clothing. Such signals, or at least some of them, must be "difficult to fake" in order to discourage the destructive infiltration of "cheaters" or what are elsewhere called "free riders"—that is, people who do not cooperate but benefit from others' willingness to do so. Humans have therefore evolved impulses which, when triggered by strong but otherwise inexplicable emotions of fellow-feeling toward others, automatically and otherwise inexplicably cause us to engage in personally costly dramatic public acts that demonstrate our sincere commitment to other members of the community; and these feelings of connectedness can themselves be triggered by the kinds of sensory stimuli (music, choral singing, bodily contact with others, and so forth) generated in communal rituals. Thus we arrive promptly at 10 am every Sunday to sing hymns and hold hands together (the example is Boyer's) and willingly and dramatically sacrifice our children, or parts of them, to the communal gods.[36]

Much of Boyer's account here, which draws on a relatively long history of research and theory in social psychology, social-exchange theory, and economic-mathematical game theory, is compelling enough, though of course limited (and, one may think, dispiriting) in the usual ways. That is, it formalizes conditions, simplifies motives, ignores individual differences and contextual mediations, and obliterates the force of, on the one hand, consciously formulated ethical principles and observed examples of ethical behavior and, on the other, the sheer perversity of human calculations. Moreover, Boyer's specific additions to the standard accounts, though novel, can be seen as gratuitous and otherwise dubious. In order, he maintains, to explain to ourselves the existence of those strong and otherwise inexplicable cooperative and altruistic impulses, we create supernatural beings with relevant information and powers of discernment who, we tell ourselves, are responsible for our moral feelings and judgments. Or, as he also puts it, because we posit gods as having "access to all relevant [strategic] information" about our own and everyone else's situations, motives, and actions, we imagine them as being able to judge what is objectively right and wrong. Summarizing the account, Boyer observes that omniscient gods are "parasitic" on our own innate impulses and capacities "in the sense that . . . successful transmission [of the idea of such gods] is greatly enhanced by mental capacities that would be there, gods or no gods."[37]

Boyer's explanations of human morality and the normative operations of religion are often, as here, highly circuitous. Indeed, many twists and turns of those explanations seem required not so much by either specific features of the behavior being explained or relevant archeological, ethnographic, or historical data as by the conceptual and methodological imperatives of the cognitive-evolutionary approach as such. That is, they are made necessary and evoked by the assumption and continuous effort to demonstrate that virtually any human behavior can be explained as the surface manifestation of an evolved capacity or underlying mental mechanism. To the extent that such imperatives operate in New Naturalist approaches to religion, they encourage myopia among its practitioners. To the extent that the resulting explanations are thin in substance and strained in argument, they prove unsatisfying to readers interested in arriving at more connectable, reliable, and illuminating understandings of how religious beliefs and practices arise, develop, and actually operate in human culture, experience, and history.

NATURALISM, OTHERWISE

Clearly one may be sympathetic to the longstanding and ongoing project of producing naturalistic accounts of religion but skeptical about various ways—past and current—of pursuing that project. The New Naturalist approaches described here are certainly promising, and many of the accounts of specific types of belief or practice developed in the name of "cognitive studies of religion" are compelling. But, as indicated above, the approaches have also tended to be limited by a number of theoretical assumptions and related methodological commitments drawn from the more general "cognitive" and "evolutionary" programs to which their practitioners give allegiance. Conversely, cognitive-evolutionary explanations of religion become more substantial, persuasive, and illuminating when they are joined to studies by scholars and researchers working with other naturalistic approaches to religion, both social scientific and humanistic, and when they incorporate other—subtler, more conceptually spacious, and arguably more empirically responsive—understandings of human behavior, cognition, and culture.[38]

The advantages of such conjunctions are demonstrated in the 1995 study *Creation of the Sacred: Tracks of Biology in Early Religions* by the distinguished German classicist Walter Burkert. Burkert's account of the origins of religious beliefs and practices, though thoroughly naturalistic, is not an example of what I have been calling the New Naturalism. Rather, in offering a series of perspectives on religion without pretensions to natural-scientific status itself, it underscores the promise of a biologically and otherwise scientifically informed approach to religion that is also informed by and connectable to broader understandings of human behavior, culture, and history. As a classical scholar, Burkert studies civilizations of the ancient world—Egypt, Mesopotamia, Israel, Greece, Rome, and so forth—as documented in literary, historiographic, and religious texts and as revealed in archeological artifacts. In drawing on such materials, he participates in a tradition of classically educated theorists of religion that includes such philosophers and social theorists as Hume, Nietzsche, and Weber as well as fellow-philologist Gomperz. Like these, Burkert is both widely literate and also exceptionally knowledgeable about relevant current work in the natural sciences, including, in his case, evolutionary biology, game theory, and biological anthropology. But *Creation of the Sacred*, originally Burkert's

Gifford lectures for 1989, was written before and without the benefit of either "evolutionary psychology" or "cognitive science" per se—and, I would suggest, without certain of their liabilities as well.

In the course of tracking biology in archaic religions, Burkert explores and analyzes primitive ideas of invisible powers and superior beings, patterns of belief and practice clustering around concepts of guilt and sacrifice, and such other familiar features of religion—modern as well as ancient—as oaths, oracles, priests, prayers, myths, and moral commandments. Like Boyer and other New Naturalists, he refers all these to general human perceptual and behavioral tendencies that are recognizable in everyday life and secular institutions; that are reflected to various extents in what we know of the capacities and responses of early hominids, primates, and other animals; and that are presumably embodied in the structure of human brains. Unlike Boyer and other New Naturalists, however, Burkert does not seek to explain any of these—either the observed religious behaviors or the inferred psychological capacities—as products of the operation of specific, discrete mental mechanisms. Moreover, while Burkert is attentive, as are the New Naturalists, to recurrent types of belief and behavior (which, for them, commonly signal universals of human nature), he is no less attentive both to the ranges of their variation under different historical and cultural conditions and also to recurrent patterns of *interrelated* concepts, behaviors, and institutions. Finally, familiar as he is with classical as well as modern languages and cultures, Burkert considers in some detail the relation between religious expression—as in myths, ritual performances, prayers, or hymns—and the more general inclinations and capacities of humans for symbolic behavior and imaginative elaboration. The title of Burkert's book is *Creation of the Sacred.* It is not "discovery" of the sacred, as Eliade and other theologically oriented scholars of religion might have preferred. But it is also not "automatic mental and emotional triggering" of the sacred, as Boyer and other New Naturalists might be inclined to put it.

A few examples from the book will illustrate these points. Burkert proposes three related ways in which ancient religions brought a sense of order and manageability to the world: first, by positing a supreme authority and hierarchical scheme of power; second, by understanding misfortune in terms of a causal pattern of crime, punishment, expiation, and salvation; and, third, by reinforcing a tendency to social reciprocity that is not only practically effective but also offers a sense of cosmic justice.[39] In a central chapter of *Creation of the Sacred* exploring religious ideas of dominance and subordination, he observes that religion "is generally accepted as a system of rank, implying dependence, subordination and submission to unseen superiors" and, in that connection, quotes the Greek dramatist Menander: "Whatever is powerful, is taken for a god." Such ideas, Burkert continues, including the familiar monotheistic concept of omnipotence, are related to ancient views of social hierarchy and honor that can be traced to more primitive relations among humans built on physical strength and height, which, in turn, can be seen as reflecting fundamental patterns of dominance relations among primates and other animals. In elaborating these observations, Burkert alludes to rank-consciousness in primate societies as described by Frans de Waal but also notes, with examples from ancient texts, that, for humans, high rank is typically associated with vertical height—as in trees, hills, mountain tops, and skies, all prominent in religious discourse—and that gods are typically "high, preferably the highest."[40] Here as throughout the book, the observations and connections abound: deftly delineated, richly illustrated, illuminating, and intellectually satisfying—though, as I note below, not for all readers.

Continuing the exploration of dominance relations, Burkert notes that, among humans, they always include some form of mutual obligation—a social-exchange contract captured by a formula that he cites from Persian inscriptions: "The Lord, honored by submission, grants protection and ensures security." Rejecting the inclination of some modern theorists to view this pattern in post-Freudian terms (for example, to explain god as "a father figure"), Burkert insists that, "if we accept an evolutionary view of anthropology, as Freud himself did," we must go back further and "account for the role of authority in *both* society and the structure of the psyche." In religion as elsewhere, he observes, power operates through a complex two-tiered structure. In submitting to god, one rises to authority among mortals; as the father bends to god, so he may expect his children to bend to him. This "religious two-tier theater of power," Burkert remarks, "still tends to manifest itself in the normal family structure." He elaborates:

> Submission and sovereignty inhabit the same hierarchic structure. Dependence on unseen powers mirrors the real [i.e. secular-political] power structure, but it is taken to be its model and to provide its legitimization. . . . In reality, while power games are played out in a continuous dialectic of aggression and anxiety, in the stabilized power structures of the human mental world this duality has become neatly dissociated, producing fear of god or gods along with constant readiness to attack and destroy lower humans, buffered by the good conscience provided by piety.[41]

The subtle and altogether unsentimental analysis here recalls both Nietzsche and Foucault: Nietzsche in the idea of a dialectics of power at the heart of social life, Foucault in the emphasis on the mutually legitimating power relations in church, state, and family. Aside from the intellectual cosmopolitanism, two other aspects of Burkert's analysis of power may be noted. First, in explaining the emergence of patterns of dominance and submission in religion, Burkert does not leap directly from observations of primate behavior and presumed Pleistocene conditions to the positing of innate, universal human mechanisms or impulses. Rather, he takes his way *overland*, tracing the many forms of social and political institution as well as psychological processes involved in the development of those patterns over the course of human history. The New Naturalists, in hopping around from one era and culture to another in order to demonstrate the generality of the religious concepts they examine and the universality of the mental mechanisms they posit, forfeit the sense of how things happen on the ground when multiple relevant forces operate together. Second, whereas New Naturalists characteristically adopt the stance of behavioral scientists and, accordingly, maintain an objectifying distance from the people whose practices and beliefs they seek to explain, Burkert represents the subjects of his study as more than just nominally fellow humans. In *Creation of the Sacred*, "they" are always—and often pointedly—"we," sharing evolutionary histories and biological impulses but also sharing recognizable desires, foibles, abilities, and achievements.

Burkert's accounts of religion are explicitly allied with sociobiology[42] and generalize without apologies across animal species, but they are not generally experienced as reductive or improperly universalizing. This is, I suspect, because they have the sort of conjoined imaginative appeal and intellectual force that derives from the elicited sense of recognition just mentioned. Yes, one says: that is the way—for better or worse—we humans are. *Creation of the Sacred* does not offer to identify the "causes," as such, of archaic religious beliefs or practices, and various of its specific biological-historical trackings can be faulted as merely speculative or empirically questionable. But Burkert does indicate compellingly how a wide array of exotic ideas and actions, described in detail, can be encompassed by recognizable frameworks of experience and observation. For this reason, that is, because his accounts of religion give the sense that remote, puzzling, and even absurd or repugnant ideas and actions fit familiar patterns of human emotion and behavior, those accounts are experienced by readers as effectively explanatory.[43]

UNDERSTANDING RITUALS

The limits of Pascal Boyer's repertoire of explanatory mechanisms in *Religion Explained* become especially evident in his treatment of religious rituals. Like other New Naturalists, Boyer not only valorizes explanation over interpretation but identifies interpretation with intellectual approaches cast as intrinsically nonscientific. Thus it is not surprising that terms like "symbol" or "represent" do not appear anywhere in his discussion of rituals or that he treats references to their "meaning" so dismissively there. Indeed, according to Boyer, rituals, contrary to the accounts of them given by many anthropologists and participants, are virtually meaningless. He writes:

> We often say that ceremonies are *meaningful* to the people who perform them. . . . This may well be what some ritual participants themselves offer as a justification for their performance. But do rituals really convey much meaning? . . . What is the information transmitted? Not much apparently. . . . [M]ost ritual language is either archaic, so no one has a clear idea of what it means, or formulaic, so that you are mandated to repeat the same words as in previous performances. . . . True, you can certainly associate various ideas with what is being done. . . . But this is mostly a matter of free interpretation, and the associations are certainly not explanations of action.[44]

This seems singularly obtuse. There are, of course, many senses of the term "meaning," but clearly the ones relevant to rituals—that is, intended and understood by participants and by most scholars of the subject—are those associated with expressive activities, in which meaningfulness is not assessed by the amount of information communicated. To speak of the meaning of a burial rite or a temple service is to speak of the ideas, emotions, connections, and reflections that its performance characteristically and more or less reliably evokes in participants—not "free" interpretations or associations, then, but relatively specific effects and responses. Moreover, it is, in part, because of the ideas, emotions, connections, and reflections that they characteristically evoke that such communal rituals are valued by humans and come to be established by and among them. Or, to put this another way: among the "causes" of communal rituals—what shape, establish, and perpetuate them—are their effects as experienced by the humans who participate in them. Such effects, which participants and anthropologists may speak of as a ritual's meaning, cannot, then, be excluded from (precisely) its "causal explanation."[45] Significantly, it is the feedback loop here—the way sociocultural institutions such as rituals are shaped and sustained by their own experienced effects—that escapes Boyer's notice and is, in fact, obscured by the strictly linear, unidirectional input-output model of human cognition that is basic to his causal theorizing.

We may recall here some common features of communal rituals: members of an otherwise dispersed group convene in an orderly manner, wear special garments, occupy special places, move together and speak or sing in unison, and view or hand around objects said to represent important ideas or important—often inaugural—events relating to the occasion. All this, each time, typically creates in participants particular effects. Notably, it heightens and organizes particular emotions, renews and enriches particular memories, enhances a sense of the specificity and weight of the occasion, and elicits feelings of connectedness both to past and future events of like kind and to other people both present and absent. Most of this is familiar to us as participants in one or another communal ritual, whether secular, religious, or quasi-religious: graduation or wedding ceremonies, Thanksgiving dinners, New Year's Eve parties, and so forth. And all of it is known in spades to scholars of religion and to other observers and analysts (archeologists, classicists, cultural anthropologists, cultural historians, and so forth) who, for the past century or two, have documented, described, compared, and interpreted communal rituals—ancient and modern, primitive and sophisticated—and sought in these ways to understand and explain them, both specifically and generally.[46]

For Boyer, however, this "intuitive" way of talking about religious rituals "cannot be the explanation." To find that, he insists, we must "do science," which means going down to "the mental basement," where we discover, once again, pre-installed, pre-programmed mechanisms and automatic outputs: hardwired inferential systems for the strategic management of social relations, precise but unconscious rational calculations of the risks and advantages of coalition, and the alleged universal syntactic rules governing the role of supernatural agents in "special actions."[47] Boyer lays out this mental machinery in some detail and illustrates its operations with a wide array of ethnographic materials. Nevertheless, it remains questionable whether his causal explanation of ritual is more empirically adequate or otherwise intellectually creditable or illuminating than the array of descriptions and interpretations produced by the anthropologists, archeologists, classicists, historians, and other scholars and theorists mentioned above—or, for that matter, more "scientific" than any of them in any rigorous sense of the term. What Boyer has produced is an additional package of speculations, at another level of description, in relation to other theoretical frameworks, in a different conceptual idiom. There are gains and losses here. What is gained is the possibility of incorporating certain aspects of religious ritual into what is currently theorized about human cognition and the evolution of social behavior more generally. What is lost—at least if Boyer's own judgments and injunctions are heeded—is the rest of the entire history of thought about the subject. Happily, we need not take the benefits on his terms.

To summarize and conclude, then: We may accept from the New Naturalists the idea that the existence, persistence, and many recurrent features of the beliefs and practices that Western scholars currently assemble under the term "religion" reflect the operation of evolved cognitive capacities and behavioral tendencies that emerge more or less reliably among humans. In accepting that idea, however, we need not accept the specific evolutionary scenarios or specific mental machinery currently posited in evolutionary psychology or mainstream cognitive science to account for human behavior or cognition more generally. Nor need we think that all of what is currently assembled under the term "religion" has thereby been explained, or that it could be, and eventually will be, explained by improved models of such machinery or improved scenarios of that kind. Nor need we think that such explanations have an exclusive claim to scientific status, or that scientificity is a sufficient dimension, or should be the sole criterion, for assessing the adequacy of an account of religion—or of anything else.

CHAPTER THREE

"The Gods Seem Here to Stay": Naturalism, Rationalism, and the Persistence of Belief

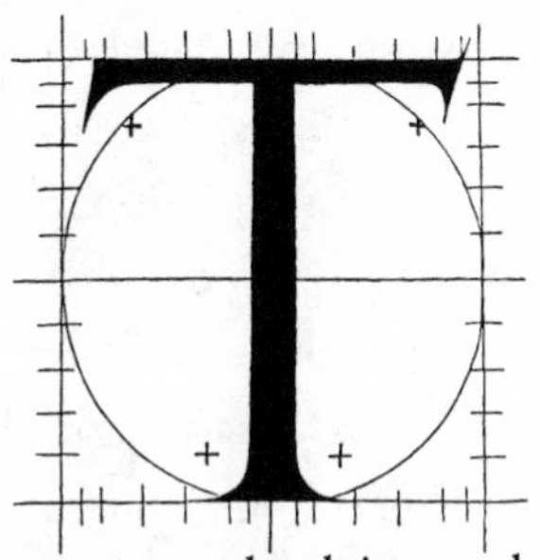

The title of Scott Atran's book *In Gods We Trust: The Evolutionary Landscape of Religion* plays, of course, on the inscription on United States coins, but in two ways: first, by pluralizing "God," which signals a not entirely reverent attitude toward the motto or the deity; and second, by suggesting that matters of finance or economics will figure in the study, as indeed they do. Atran opens the book by posing what he calls "an evolutionary riddle," namely: what maintains religious belief among humans in view of its apparent costs?[1] What makes this an *evolutionary* riddle, he explains, is that the costs involved—unreciprocated sacrifices to supernatural beings, wasteful expenditures of resources in ritual displays, and so forth—are effects of religious belief that would seem to reduce biological fitness. The answer to the riddle, he goes on to suggest, here in line with New Naturalist thought more generally, is that many beliefs associated with religion reflect the operation of innate cognitive systems that were naturally selected because they secured fitness benefits for our Paleolithic ancestors. But, Atran continues, here more distinctively, many of those beliefs and related practices also meet enduring emotional and social needs of humans, thereby yielding important benefits and making religion a powerful and perhaps irreplaceable cultural system, unlikely to be superseded by science or any secular ideology. The claim is clearly significant for contemporary thinking about religion and secularization.

Among other notable features of Atran's subsequent assessment of the biological "payoffs" of religion is his even-handedness in considering the more proximate effects involved. Thus, observing that religion is not a single evolutionary entity but the product of a variety of cognitive systems, he continues: "Even those systems with an evolutionary history . . . have often been culturally 'exapted' to new functions absent from ancestral environments. And these may have little, if any, systematic relation to genetic fitness: encouraging xenophobia or opposing injustice, supporting ruling cliques or their overthrow, facilitating social communion or con games, relieving or inciting anxiety, enhancing or suppressing sexual pleasure, promoting or impeding artistic creativity, fostering or hindering insight into ourselves and the world around us, seeking or inventing the truth, and the like." This list of the "functions" of the cognitive systems involved in religion is clearly painstaking in its effort at balance, with each item from the familiar catalogue of the benefits of religion matched against an item from the familiar catalogue of its crimes.[2] To be sure, the crimes are rather diplomatically named here and Atran has some difficulty maintaining the balance throughout the book. (As discussed later in this chapter, he ignores or soft-pedals many features of religions that would give added bite to the debit side of his accounting sheets or complicate his cost-benefit calculations.) Nevertheless, Atran's emphasis on the ambivalent—positive and negative—operations of religion ("for better or worse," as he writes in a key passage later)[3] is valuable as such and is one of the many merits of the book. Indeed, in its intellectual scope and the precision of its formulations, *In Gods We Trust* is a showcase of current New Naturalist—that is, cognitive-evolutionary—approaches to the study of religion. At the same time and often for those very reasons, Atran's book demonstrates with special clarity some of the characteristic limits and liabilities of those approaches.

Using Atran's claims and arguments as examples or points of entry, I continue in this chapter the examination of New Naturalist accounts of religion begun in chapter 2. My focus here is, first, the intellectual costs of Atran's commitment to dubious models and methods drawn from evolutionary psychology and, second, a persistent rationalist bias in New Naturalist understandings of the concept of belief—here with additional examples from Daniel Dennett, *Breaking the Spell: Religion as a Natural Phenomenon.* In the final sections, I consider Atran's and other New Naturalists' accounts of the biological benefits of religious belief and draw out some of the implications of those accounts for thinking about the future of what we call "religion."

NEW NATURALIST COMMITMENTS

As indicated in chapter 2, the group of theorists I am calling here the New Naturalists look mainly to evolutionary biology and cognitive science for their accounts of religion. Largely social scientists themselves, they apply evolutionary biology to human behavior and culture primarily in accord with models, theories, and methods developed in such relatively recent fields as evolutionary psychology, rational-choice game theory, and cognitive anthropology. To the extent that New Naturalist accounts of religion are thereby derived, they reflect the specific programmatic assumptions that define these fields and, with them, various conceptual biases, methodological confinements, and ideological taboos.

For Pascal Boyer, Atran, and most other New Naturalists, the notion of an adapted computational-modular mind operates as intellectual bedrock, with the usual sort of mixed consequences. On the one hand, it energizes their efforts and gives a show of theoretical coherence to their explanations of many features of religion. On the other hand, it tends to lock their thinking into a rigid set of ideas and conceptual moves that limits both what they notice about religion and what they take into account as relevant to its explanation. The myriad phenomena that constitute religion tend to be cut down to modular size; whatever cannot be explained by an adapted mental module tends to disappear as a phenomenon. Anthropologists forget the concrete particulars of their field work; psychologists forget their lifelong observations of the individual quirks of human behavior; researchers discount the empirical studies and theoretical achievements of predecessors and of contemporaries working in related fields; scholars ignore the archives of human history.

One consequence of this set of commitments is an eagerness on the part of New Naturalists to posit evolved mental mechanisms to account for all general and/or recurrent features of human behavior and culture. Although Atran correctly cautions against evolutionary scenarios based on minimal archeological and paleoanthropological evidence,[4] he nevertheless participates in this eagerness in *In Gods We Trust*. Thus, while he observes early on that, given the variety of phenomena involved in religion, we should not seek to explain everything about it as the product of innate cognitive mechanisms, one of his major claims is close to the reverse of that. Where we do see a high degree of similarity among the beliefs and practices of people living in different cultures, Atran argues, we may confidently posit the existence and operation of just such mechanisms. Such extensive cross-cultural convergence, he writes, is "possible only because of the universal structural framework" of the mind, which "involves innate and modularized expectations about object movements . . . , essential kinds . . . , the intentional nature of agents," and so forth.[5] Atran's leap here from the observation of similarities in the beliefs and practices of different cultures to the confident positing of universal mental mechanisms is a key conceptual move in evolutionary psychology and many New Naturalist explanations of religion but, as here, a dubious one.

Expanding his claim of the significant constraining effects of presumed universal mechanisms, Atran adduces as evidence the resemblance of the Egyptian phoenix to the pre-Columbian feathered serpent and comparable similarities between such other imaginary creatures as Irish elves and the fairies of Russian folklore. He writes: "If we reject the unlikely possibility that these thematic recurrences stem from historical contact and diffusion or are spontaneous instantiations of a Platonistic set of innate religious forms . . . , then how else could such apparent recurrences independently take place across cultures without specific and strong universal cognitive constraints?"[6] The answer to Atran's rhetorical question here—*how else?*—is not, as he implies, *no other way*, for these are not the only alternatives. Rather, like the virtually universal metaphors of light and dark or battles and journeys examined by George Lakoff and Mark Johnson in their classic study *Metaphors We Live By*, the recurrent content of folk beliefs could be explained—without invoking cultural diffusion, Platonistic forms, *or* "specific and strong universal cognitive constraints"—as the product of recurrent human activities, impulses, and experiences under widespread human conditions. The sun rises and sets cross-culturally; people everywhere go on journeys; snakes and birds are found together in many regions of the globe. The recurrent construction of chimera—imaginary creatures combining parts and attributes of known ones—no doubt involves the operation of general human tendencies, cognitive and other. But the appearance of such creatures in the lore of remote and otherwise very different cultures does not require the positing of highly specialized innate mental mechanisms for their explanation. Nor does the recurrence, stability, or cross-cultural similarities of these and other "symbolic" beliefs constitute evidence for the fundamentally constraining operation of such mechanisms.[7]

The dubious conceptual moves involved in the evolutionary-psychological/New Naturalist generation of mental mechanisms are illustrated in another passage in Atran's *In Gods We Trust*. Here, associating our alleged intuitive comprehension of the concept of gods with children's alleged intuitive knowledge of grammar, he writes: "Without such biologically poised competencies—the products of millions of years of biological and cognitive evolution—a child's acquisition of such a wide range of cultural knowledge in a few short years would be miraculous. This is also true for any adult's ability to network such a vast repository of detailed information in a mere lifetime, or for any anthropologist's ability to understand something of an alien culture in a season or two of fieldwork."[8] The argument, familiar from Noam Chomsky's theory of language acquisition and repeated by other New Naturalists, is that, to avoid believing in miracles of human learning, we must believe in a storehouse of prior inner competencies—the existence of which, for many critics of the argument, would be hardly less miraculous.[9] Contrary to Atran's suggestion, explanations of the phenomena at issue are by no means impossible without the assumption of such inner competencies. An unshakable conviction of the existence of such competencies, however, is likely to make alternative explanations more difficult to discover or recognize. Atran's examples here of supposedly otherwise inexplicable human achievements—learning to speak one's native language, mastering the basic informational repository of one's native culture, gaining insight into other cultures—are highly complex and individually quite different types of accomplishment, each the endpoint of an array of processes that occur over time. An adequate explanation of any one of them would seem to require, at the least, consideration, observation, and analysis of the types of experiences, developmental processes, and cultural resources that might be involved in their attainment. Explanations of such complex human behaviors in terms of innate inner competencies, however, make such empirical and analytic activities unnecessary—or, to put it the other way around, foreclose such activities before they can begin.

We might note in passing that, while Atran and other New Naturalists typically stress, as in the passage above, the "millions of years" involved in the evolution of hypothetical cognitive mechanisms, they also typically neglect the many years involved in the maturation of every human being—what Atran refers to here as "a mere lifetime." As suggested by (among other things) the shelvesful of memoirs and testimonies from Augustine's *Confessions* to contemporary tales of personal enlightenment or reversion, if one's aim is to understand and explain the religious beliefs and behaviors of human beings, an individual lifetime is a not insignificant unit.

In seeking to account for any complex behavioral, cultural, or social phenomenon, a good starting assumption would be that it was the emergent outcome of multiple factors of various kinds, operating at many scales and levels, interacting over time. The starting assumption of evolutionary psychology and "cognitive" approaches to religion, however, is that the best way to explain any behavioral, cultural, or social phenomenon is by demonstrating that it is the outward effect of the activation of some underlying mental mechanism. A methodological tradition of this sort puts a premium on ingenuity with respect to the hypothesizing of mental mechanisms and, by the same token, encourages negligence with respect to the investigation of possibly relevant environmental, experiential, and developmental factors. Such a tradition also encourages a readiness to dismiss, as always already discredited, alternative explanations in which such factors figure significantly. Like Boyer and other New Naturalists, Atran frequently draws comparisons between his own cognitive-evolutionary account of some feature of religion and other existing psychological or sociological accounts, with the alternatives posed as in zero-sum competition and the cognitive-evolutionary account presented as superseding any and all others. In many cases, however, the strenuously rejected alternative could just as readily be seen as an additional—complementary and compatible—explanation of a complex phenomenon that is the product of multiple interacting causes.

For example, in explaining the origin and attributes of supernatural agents, Atran discusses, as a "competing theory" to the cognitive-evolutionary account in terms of an evolved innate module for agent-detection, the account offered by so-called Attachment Theory. According to the latter theory, many common features of gods and religious worship reflect people's extension of their childhood experiences of and attitudes toward parents. With respect to gods, such features would include benevolence, power, concern for worshippers' welfare, and the reward and punishment of worshippers' behavior. With respect to worship, they would include attitudes of veneration, dependence, and the desire to please and/or appease. Attachment Theory, at least in the versions Atran paraphrases and discusses, is clearly inadequate as a comprehensive explanation of the origin of deities: gods (and parents) are not always benevolent and many other common attributes of deities, especially in polytheistic systems, would remain to be explained. But Atran's lengthy and laborious refutation of Attachment Theory's most dubious claims does not clean the slate—as he clearly intends—of all past, current, or possible notions of a connection between concepts of gods and childhood experiences of parents or elders.[10] Moreover, the proposal of such a connection is not inherently, as Atran represents it, a rival to the cognitive explanation of such concepts in terms of a hyperactive agent-detection device. It could just as readily be seen as a complementary account. The parent-like features of gods at issue (benevolence, concern for worshippers' welfare, and so forth) could be understood as projected onto the supernatural agents presumably thus created in pretty much the same way as other anthropomorphic features are said to be "bundled" onto them in Atran's own account. ("Once a source is identified as an agent it is bundled with all sorts of attributes, which usually identify an animate agent instigator.")[11] At the same time, the worshippers' childlike attitudes (dependence, veneration, the desire to please and/or appease, and so forth) could be seen as similarly shaped by multiple factors. These might include the sorts of experiential factors proposed by Attachment Theory, the sorts of evolved cognitive mechanisms proposed by Atran and other New Naturalists, and the broadly cultural and historical as well as biological factors proposed by Walter Burkert in his analysis of the hierarchical power relations between deities and worshippers found in many religions.

Several reasons may be offered for the New Naturalists' typical duel-to-the-death stance toward alternative—possible or existing—explanations of various features of religion. A general resistance to alternative views is an observable effect, of course, of most established belief-systems and is often especially intense among advocates of innovative intellectual positions or under conditions of ongoing acute controversy. But an additional configuration of pressures and biases seems to be operating here. First, as already suggested, if one is persuaded that explanations of human behavior in terms of evolved hardwired cognitive mechanisms are more truly explanatory than any other type, then as soon as one has devised a minimally plausible cognitive-evolutionary account of some behavioral phenomenon, one will have no intellectual incentive to consider what other factors may be involved. If, moreover, the explanatory invocation of a range of other factors (here, social interactions, cultural conditions, and/or individual experiences) is associated with one or more communal taboo or supposed intellectual heresy or naivety (such as, here, "social constructionism," "cultural relativism," or an alleged "blank slate" model of mind),[12] then a positive disincentive will be added to such pressures. A significant consequence of these communal biases, taboos, and heresy-aversions among New Naturalists is the obliteration from their explanatory repertoire of a very wide range of important resources for understanding religion.

RATIONALIST BELIEFS

One of the most characteristic but in some ways puzzling features of New Naturalist accounts of religion is a set of presumptions about the nature of belief associated with traditional rationalism. By "rationalism" I mean here the system of formal and informal views (doctrines, assumptions, definitions, ascriptions, and so forth) that attend the idea of a distinctively human faculty, *reason*, the proper exercise of which, understood as due conformity to supposedly objective rules of logical inference, is thought to direct us to beliefs, attitudes, and actions that are correct, sensible, and appropriate. Rationalism in this sense presupposes a specific partitioning of human faculties (reason, imagination, the will, the passions, and so forth) that is hard to square with current neuroscientific understandings of brain anatomy or function; and it is accompanied by a network of traditional, often very ancient, ideas and related distinctions (brute facts, mere fancies, direct perceptions, a universal common sense, and so forth) that are comparably questionable from the perspective of much contemporary epistemology and empirical psychology.[13] In short, rationalism is a rather old-fashioned view for a group of researchers and theorists who present themselves otherwise as intellectual innovators and iconoclasts. In many significant ways, however, and often quite explicitly, the New Naturalism is a direct descendent of the Old Enlightenment.[14] Insofar as the resulting legacy involves dubious rationalist views of cognition, including a conception of beliefs as comprised of discrete true/false, rationally justified/irrational propositions about, or correct/incorrect mental representations of, an autonomous reality, it is distinctly limiting when it comes to understanding the nature and operations of what we call religion.

A rationalist bias reveals itself early on in Atran's book in his elaboration of the New Naturalist idea that concepts of gods or supernatural beings are especially compelling to the human mind because they involve a conjunction of intuitive and counterintuitive attributes.[15] I comment on various problematic features of this idea below. Of interest here, however, is Atran's claim, in elaborating it, that people everywhere recognize "a principled distinction between intuitive and counterintuitive beliefs." There must be such a distinction, he observes, because, if "commonsense" beliefs about "brute facts" were processed the same way as "symbolic" (counterfactual or imaginative) beliefs, then the commonsense beliefs "would invariably contradict" the symbolic beliefs and presumably cancel them out. Moreover, he continues, if there were no such principled distinction, "any conclusion would logically follow from any belief because people would assent to blatantly contradictory facts."[16] As I think is clear, the sharp distinction between two types of belief is required, and the claims and arguments for that distinction make sense, only on the rationalist and logicist assumption that we cannot harbor contradictory beliefs and do not assent to "blatantly contradictory facts."

Several points may be made here. First, as noted in chapter 1 above, the term "belief" embraces a host of quite varied phenomena ranging from discrete, verbally framed creeds to vague mental images and more or less general ongoing behavioral/perceptual dispositions. The division of that experientially and physiologically heterogeneous array into two distinct and differentially processed types of belief appears quite arbitrary and, as part of Atran's argument for the particular double constitution of supernatural concepts (intuitive and counterintuitive), question-begging. No less problematic is the idea of a universally recognized distinction between, on the one hand, commonsense beliefs about brute facts and, on the other hand, symbolic beliefs about supernatural entities. To be sure, the operation of some form of psychological compartmentalization is suggested by such otherwise paradoxical- (or hypocritical-) seeming phenomena as the religious devotion of evolutionary biologists or the thrills experienced by atheist fans of ghost stories. But, rather than posit discrete innate mental categories and/or sharply distinctive types of cognitive processing to account for them, we could see such phenomena as evidence of the continuously shifting configuration of our ideas and impulses and their context-sensitive weighting and emergence. What Atran seems to want to maintain here is the existence of a physiological difference—a distinction and segregation in the very structure and operation of our brains—between, on the one hand, factual knowledge and veridical intuitions about the world and, on the other hand, counterfactual concepts and other merely mental constructs. There is, however, little to support such an idea aside from rationalist prejudice.

A rationalist (and, here, realist) epistemology is explicit in Atran's discussion of so-called neurotheology. Neuroscientists Eugene d'Aquili and Andrew Newberg claim to have demonstrated, on the basis of brain-imaging studies of meditating Buddhists and nuns in prayer, that the experience of God (what they term "Absolute Unitary Being") is hardwired into the human brain.[17] Atran takes particular issue with d'Aquili and Newberg's observation that "the experience of God . . . is as 'real' as the experience of ordinary objects and events." He also objects strenuously to their related observation that (as Atran paraphrases it) "science cannot decide whether ordinary, baseline perception and conception or the mystical perception and conception of God is 'more real.'" He comments with some energy: "This is a specious argument. What evolutionary science *can* tell us is that without our ordinary, baseline reality our species would have had no natural means of emerging from the evolutionary process. If our ordinary perceptions and inferences were not at least approximately veridical with respect to everyday objects and events, then how . . . could our ancestors have possibly avoided falling off cliffs or being eaten by wild animals at every turn?"[18]

One may share Atran's skepticism about the theo-ontological conclusions drawn from brain-imaging studies without accepting the idea that our ancestors' survival of prehistoric cliffs and predators is evidence for the normative ("baseline") veridicality of our "ordinary" perceptions and reasoning. For one thing, the notion of "ordinary" human perceptions—presumably meaning independent of drugs, dreams, fevers, anxieties, obsessions, or astigmatisms—is itself problematic and, again, question-begging here. Everything we know about human psychology indicates that both what we perceive and how we respond to our perceptions are shaped by multiple contextual and individuating factors, including more or less transient physical and emotional states and specific past and ongoing social, cultural, and, as may happen, "religious" experiences. More significantly here, the same epistemological argument that Atran makes for ordinary human perceptions and conceptions could be made for the ordinary perceptions and conceptions of bats, bees, frogs, and every other creature whose lineage has survived the particular catastrophes to which its kind is vulnerable. The latter view, that all extant creatures generally see what is really there, can of course be maintained, but only if one also accepts the idea of multiple realities. This is, however, an idea that Atran specifically rejects and seeks here, as elsewhere in his book, to refute. Thus, discussing such practices as voodoo or tribal healing, he insists (contra "relativists") that the effectiveness of such practices for members of tribal communities does not indicate an "alternative reality" that has "no rational explanation." Rather, he writes, such practices "partake of panhuman cognitive principles applied to a different ecological context."[19] Atran might maintain that there are also panfrog and panbat cognitive principles, but the question of the epistemological status of, and ontological relations among, the resulting frog, bat, and human realities would remain. Moreover, once one acknowledges the possibility of contextual variability in the perception of realities, as Atran implicitly does here (and as, I think, one always sooner or later must), then the normative force of those supposedly pan*human* cognitive principles—their value, that is, for distinguishing ("rationally") between the really real and the really not-real (for example, the merely mystical)—has been crucially compromised.

A different view of the relation among diverse experiential realities—and also between human cognitive products and those of other creatures—is available and, I would suggest, preferable here. In accord with the constructivist-pragmatist epistemology outlined in chapter 1 above, the veridicality of any creature's, including any individual human's, cognitive processes can be seen not as the accuracy of its perceptions of a presumptively objective reality ("what's really there") but as the relative effectiveness of that creature's ongoing interactions with its particular environment, given its particular structure and modes of operation. The "real," under such a conception, would be understood as the more or less stable and more or less pragmatically workable cognitive constructs produced by that creature through those more or less effective interactions—which, in the case of humans, would include verbal interactions with other humans and the effects of ongoing and past experiences of the cultural and social environment more generally. The processes and products of "mystical" experiences can be understood, then, as only more or less different from the processes and products of "ordinary, everyday" experiences. The sharpness of difference between them, and also the specific interpretations and imaginative elaborations of the constructs produced by each, would themselves be seen as depending crucially on local contextual factors that ranged from culturally, historically, and institutionally dominant discourses (for example, discourses involving angels, visions, and transports versus those involving modules, hardwiring, and neural firings) to matters of highly individuated personal temperament and personal history.

Finally here, with regard to Atran's convictions concerning the difficulty or impossibility of harboring contradictory beliefs, we may return to the idea, introduced in chapter 1 above, that the sets of beliefs held by each of us are fundamentally incoherent—that is, heterogeneous, fragmentary, and, though often viable enough in specific contexts, potentially logically conflicting. We recall the unusual collection of ideas that Marion Keech and her followers in some sense believed, which included communication via automatic writing, visits to earth by superior beings from another planet, and the imminence of both an apocalyptic geological catastrophe and their own escape and salvation through faith. As I noted earlier, although the heterogeneity and incoherence are especially dramatic here, comparably mixed, variable, and fragmentary cognitive assemblages have been documented and theorized by numerous scholars. Thus Catherine Bell, an ethnographer and scholar of Asian religions, observes that "belief in spirits" among the Chinese involves widely varying types of conviction within individuals as well as among them and concludes accordingly: "There is little to suggest that a belief in spirits . . . is any one sort of belief. There is, in other words, very little systematic coherence." Similarly, classicist Paul Veyne, in his book *Did the Greeks Believe in Their Myths?*, concludes from his studies of ancient thought and culture that an adequate answer to that question cannot be framed in terms of such modern distinctions as "real" versus "imagined" or "true" versus "fictional."[20] With regard to contemporary Americans, professor of religion Robert Orsi observes bemusedly that some of his students from Bible-reading homes, rather than rejecting their parents' beliefs, "have put together intricate Christian understandings that draw on neo-paganism, snippets of Asian religions, popular psychology, and contemporary science fiction."[21] And, of course, we are all aware of the diverse array of ideas and dispositions that we carry around in our own heads (and bodies): creedal statements learned in childhood, emotion-laden memories and habits, academically acquired knowledge, individually worked-out convictions that vary in strength and articulateness from one context to another, vagrant images, transient impulses, and so forth. In the face of such evidence of the fluidity, variability, and heterogeneity of cognitive states, cognitive processes, and mental content-types, the continued invocation and deployment of static, atomistic, logicist, and dualistic conceptions of belief by philosophers of mind and cognitive scientists is itself a revealing example of the peculiar (and officially irrational) operations of human cognition.

Rationalist conceptions of belief are especially hobbling when it comes to understanding and assessing the operations of belief in the lives—experiential, imaginative, emotional, and practical—of religious people. In that connection, we may look briefly at Daniel Dennett's recent book *Breaking the Spell*, which promotes naturalistic explanations of religion, summarizes the key ideas of a number of New Naturalists, and entertains the idea of scientifically informed social policies aimed at the quarantine and eventual eradication of "toxic" religious convictions.[22]

In a central chapter of the book titled "Belief in Belief,"

Dennett, well known as a philosopher of mind and promoter of Darwinian theory, undertakes an investigation into the question "whether we have *good reasons* for believing in God." In the course of the investigation, he describes and contrasts various types of beliefs, all characterized as discrete mental propositions or distinct mental states. Following a standard analytic account of the definitive or conceptually proper relations among *beliefs*, *actions*, and *understanding*, Dennett launches into a formal comparison of religious beliefs with what he refers to as beliefs based on facts and beliefs about scientific truths. His key claim here is that, while many modern, educated people believe that religious beliefs are a good thing to have (thus "belief in belief") and may even pay "lip service" to such beliefs themselves, they do not believe them in the same way as they believe what they actually *understand.*[23] He writes: "What is commonly referred to as 'religious belief' or 'religious conviction' might less misleadingly be called *religious professing.* Unlike academic professors, religious professors [that is, people who "profess" religious beliefs] . . . may not either understand or believe what they are professing." In support of the claim, Dennett observes that if people genuinely and properly *believed* the religious propositions they professed, they would not hesitate to subject them to empirical tests, as scientists do to their own theories. He comments pointedly: "And this makes a huge difference because it gives beliefs about the truths of physics a place where the rubber meets the road, where there is more than professing that can be done. . . . [T]hese are beliefs that you can act on in ways that speak louder than words."[24]

It seems clear from the terms of the contrasts here—between beliefs you can act on and just words, between things really done and mere professing—that Dennett is unaware of what, in his terms, could be called the simplest relevant empirical facts. As documented by legions of scholars of religion drawing on observations gathered across the globe, the overt or covert verbal professing of creedal propositions is rarely the center and certainly not the whole of religion for most believers, simple or sophisticated. On the contrary, people who identify themselves as religious, including educated people in apparently secularized societies, are likely to organize and conduct a good part of their lives around the substantive, bodily acting out of their religious identities in highly consequential, socially and emotionally risky, materially costly ways—precisely, one could say, where the rubber meets the road.[25] Moreover, as emphasized by naturalists both old and new, even those religious activities that may appear most like mere professing, such as participating in liturgical services, have crucial emotional and social effects on—or, as it might be seen, perform crucial functions for—participants and their communities. Given Dennett's strictly above-the-neck, disembodied conception of belief, he has no way of registering those activities, effects, or functions, much less understanding the complex, multileveled forces that shape and sustain them.

Central to Dennett's conception of religious belief and recalling Atran's comparable dichotomies (discussed above) is what Dennett represents as two fundamentally distinct and contrasting mental states. Thus, after rehearsing standard academic-philosophical distinctions between facts and fictions, or "real" things and "things you can [only] think about," he turns to a main point of the chapter, the difference between "belief in belief" and "belief in God." He writes of them:

"[T]here is a clear empirical difference between these two states of mind."[26] The notion of an *empirical* difference between states of mind, no less a "clear" one, must give pause to a good many of Dennett's fellow philosophers. More significant here, however, is the way these supposedly sharp distinctions confine his views of the operations and effects—social and political as well as individual and cognitive—of religion.

In the concluding chapter of *Breaking the Spell*, Dennett lays out a program of desirable further scientific research into religious belief and, as a key point in the program, the need for scientists to develop "specific [and 'testable'] hypotheses about patterns in human tendencies to respond to religion." (As he indicates, Dennett himself favors the hypothesis, not itself obviously testable, that religious beliefs operate as culturally transmitted viral "memes.") Hypotheses about such patterns, he observes, would "prove useful in disentangling some of the vexing policy questions that we have to face." He continues: "[I]t would be particularly useful to know more about how secular beliefs differ from religious beliefs (and as we saw . . . , 'belief' is a misnomer here; we might better call them religious *convictions* to mark the difference). How do religious convictions differ from secular beliefs in the manner of their acquisition, persistence, and extinction, and in the roles they play in people's motivation and behavior?"[27] Especially notable in this tendentiously framed program of scientific research, with its exclusive focus on the detailing of already presumed differences, is Dennett's neglect of the possibility of any relevant relations of similarity, interaction, or continuity between religious convictions and secular (for example, social and political) ones. For what he fails, accordingly, to consider is the usefulness, for "the policy questions we have to face" and otherwise, of observing how the two sorts of conviction do, in fact, relate to and resemble each other—precisely, among other ways, "in the manner of their acquisition, persistence, and extinction, and in the roles they play in people's motivation and behavior."[28]

BIOLOGICAL ECONOMICS AND RELIGIOUS BELIEF

New Naturalist accounts and assessments of religion have implications, of course, for thinking about its likely future, and Atran, among others, has spelled some of them out. Like others who consider the question, he is especially concerned in *In Gods We Trust* with how religion (identified with belief in and worship of supernatural entities) is likely to fare in competition with "science" (here, as commonly elsewhere in these discussions, presumed to need no definition). A number of practitioners and promoters of the New Naturalism endorse a fairly straightforward secularization thesis, seeing traditional religious belief disappearing with the spread of scientific knowledge and especially with the spread of naturalistic explanations of religion. Other New Naturalists stress the deep attractiveness of supernatural concepts for the human mind and warn of the fragility of science in competition with religion.[29] Atran, fairly distinctively in this regard though arguing from comparable evolutionary-cognitive assumptions, stresses the important emotional, social, and ecological functions that religion serves and sees it as at least holding its own with science into the foreseeable future. Thus he writes at the end of his book: "[N]o other mode of thought and behavior deals routinely and comprehensively with the moral and existential dilemmas that panhuman emotions and cognitions force on human awareness and social life. . . . As long as people share hope beyond reason, religion will persevere. For better or worse, religious belief in the supernatural seems here to stay."[30]

In tallying the benefits of religion, Atran emphasizes its satisfaction of what he sees as certain fundamental emotional needs of individuals and also its ability—unique, he believes, here in company with many theistic commentators—to sustain the moral norms of communities against otherwise fatally destructive forces. Among the emotional needs is what he refers to as a universal "existential dread of death." As Atran explains it, a faculty for reasoning that originally evolved in humans to enable our ancestors to make inferences about situations relevant to their physical survival led eventually to their awareness of the inevitability of their own death and that of persons close to them. He calls this "the Tragedy of Cognition." But, his account continues, our ancestors alleviated the burden of this awareness by inventing a supernatural world and converting human death from an inevitable and unhappy event into a "telic" one whose "goal state" is "an extended afterlife."[31] Atran concludes the account by observing that secular ideologies offer little that can compare to the comfort provided by the concepts of personal immortality and an afterlife and that science in particular offers only the cold comfort of knowing that we are all in the same boat.

This account of the origin of concepts of immortality and an afterlife is not altogether unfamiliar. It recalls, among other things, the "emotive" theory of religious concepts that Boyer characterizes in *Religion Explained* as intuitively appealing but not up to par scientifically. Indeed, it is not clear that Atran's account here is anything other than that theory with the central emotion involved—what Boyer refers to dismissively as a supposed "abstract metaphysical dread of mortality"—now given a duly "cognitive" explanation as the output of an evolved mental module. The same observation could be made of a good many other New Naturalist explanations, in which familiar psychological, sociological, or anthropological accounts of various aspects of religion—rituals, priests, moral codes, and so forth—are rejected as insufficiently scientific only to be served up again in cognitive sauce.

Two further points are worth noting here. One is that, although Atran and Boyer both invoke hypothetical mental mechanisms and generate universalist cognitive-evolutionary scenarios in their accounts, they end up with widely divergent explanations of concepts of immortality and an afterlife. For Atran, such concepts may be traced to an existential dread of death arising from humans' acute consciousness of their own mortality. For Boyer, they consist of a suite of unconscious automatic responses to the presence of corpses. Evidently the explanatory products of a cognitive-evolutionary approach to religion depend on the individual theoretical tastes of whoever pursues it, which suggests that the reliability of its defining methodology is a bit shaky.

The other point of interest here is that these divergent explanations of concepts of immortality imply opposed assessments of the value of those concepts. In Boyer's account, our emotions in the presence of dead bodies are automatic and unconscious and just come to be interpreted by some people in terms of such already existing supernatural concepts as immortal souls or gods living in invisible places.[32] Together with the relatively breezy style in which it is delivered (very different from Atran's solemn evocation of an existential human tragedy), Boyer's account implies that we would do well to recognize the unconscious causes of our feelings about corpses and, if we have connected those feelings to ideas of immortal spirits dwelling with gods in heaven, to just let such connections go. For Atran, on the other hand, ideas of immortality and an afterlife are crucial in alleviating fundamental emotions of dread, grief, and despair that would otherwise overwhelm us. Such concepts, in Atran's view, are part of what has kept—and still keeps—the species going.

In addition to seeing religious concepts as providing such emotional benefits, Atran locates their continuing biological value in their power to control human deceitfulness and mitigate ruthless egotism. Without such concepts as watchful gods, he suggests, we would have a brutish Hobbesian jungle: "Social competition virtually guarantees the exercise of deception. . . . [I]f people cannot believe that the moral basis of community life is directed by the fundamental nature of the world in an intelligent, purposeful, and meaningfully designed way, there is nothing except brute force to prevent defection and nothing except individual goodwill to prevent deception." Moreover, Atran is persuaded that if and when people discover these truths about human motives, their moral commitments grow weak:

> If people learn that all apparent commitment is self-interested convenience or, worse, manipulation for the self-interest of others, then their commitment is debased and withers. In times of vulnerability and stress, social deception and defection in the pursuit of self-preservation are more likely to occur.
>
> . . .
>
> To keep the morally corrosive temptations to deceive or defect under control, *all concerned*—whether beggar *or* king—must truly believe that the gods are always watching.[33]

These observations are in good accord with New Naturalist notions of how individual self-interest, evolved social stratagems, and unconscious game-theoretical calculations figure in the generation of various religious and social practices. What remains unclear, however, is how, in all this, Atran accounts for his own evidently decent behavior or how he would account for the decent behavior, such as it may be, of his New Naturalist and other secularist colleagues, many of whom have expressed their confidence that there are no gods watching. My point is not that they must all be secret believers or nasty people. It is, rather, that we need a better account than Atran—or the New Naturalism generally—seems able to provide of what keeps most people, nonbelievers included, minimally ethical. An account of that kind might also be expected to tell us what makes—or has made—some people, nonbelievers included, maximally so.

A section of *In Gods We Trust* concerned with the communal benefits of religion—here, ecological ones—reflects Atran's extensive experience as an anthropologist and illustrates (though not intentionally) the strengths of more traditional cultural-anthropological approaches to the study of collective beliefs.[34] He is especially concerned here to refute the idea of an inevitable "tragedy of the commons," particularly as represented by Jared Diamond's suggestion that the devastation of Easter Island was the product of the environment-destroying and ultimately suicidal beliefs and rituals of its inhabitants.[35] In explicit contrast to such an idea, Atran proposes an environment- and self-preserving "spirit of the commons." His argument, with contemporary field work on the Itza' Mayans as ethnographic evidence, is that the projection of an intentional reciprocity between humans and "spirits" can operate to conserve communal natural resources where evolutionary pressures toward individual self-interest alone might otherwise lead to destructive practices. For example, he notes, communal concerns about crops and animals transmitted in local myth and folklore as the wishes or punishments of the spirits may embody summaries of empirical observations of interactions between humans and other species made by community members over many generations. More generally, Atran observes, members of communities with such beliefs tend to regard nonhumans, from fish to forests, not simply as material resources or potential commodities but, in apparent defiance of how costs and benefits are commonly figured in game-theoretical rational-choice models of human behavior, as fellow "players" in the same ongoing life-game as humans.[36]

It is a compelling set of observations. Although the familiar cognitive tendencies Atran invokes here—anthropomorphism, animism, projection, and so forth—are characterized, in his discussion, largely in terms of their supposed evolutionary histories, his analysis on this occasion departs from standard New Naturalist practice in a number of significant ways. One is in giving something like due consideration to the effects of both experiential learning (via the "mobiliz[ation]" of "inferential processes," in his lingo) and cultural resources. Another is in investigating closely the interactive relations between humans and their continuing and current environments, not just speculating about likely "inputs" and responses in supposed prehistoric environments. A third, which returns us to a point made earlier, is that Atran focuses here on the dynamics of those interactive relations over historical, not just evolutionary, time—"many generations," not the usual "millions of years."

Illuminating interludes of this kind are among the intellectual attractions of *In Gods We Trust*, but the mental-module "cognitive" accounts, along with Atran's more general commitment to paradigmatic evolutionary psychology, remain central. In the concluding pages of the book, reiterating and reflecting on the various individual and communal benefits of religion that he has traced there, Atran writes: "All of this isn't to say that *the* function of religion is to neutralize moral relativity and establish social order[,] . . . to promise resolution of all outstanding existential anxieties, . . . or to explain the unobservable origins of things, and so forth. Religion has no evolutionary function per se." Nevertheless, he continues, "the cognitive invention, cultural selection, and historical survival of religious beliefs owes, in part, to [their] success in accommodating" the "moral sentiments and existential anxieties" that "constitute—by virtue of evolution—ineluctable elements of the human condition." Or, as he writes in another summary section: "[R]eligions in general—and gods, ghosts, devils, and demons in particular—are culturally ubiquitous because they invariably meet or systematically manipulate modular input conditions of the human mind/brain."[37] In short, for Atran, what finally and fundamentally explains the origin, ubiquity, and endurance of the gods are uniform features of "the human mind/brain" as produced by evolution and as specifically described and theorized in evolutionary psychology.

The resulting focus on biology, prehistory, and what is posited as the adapted modular mind, along with the programmatic rejection of the social as a domain either of emergent phenomena or of causal dynamics, obscures more than a century of relevant work in social theory and the sociology of religion. Although Atran cites Durkheim and Weber for individual ideas in discussing the social functions of religion, he makes no evident use of their insights into the ways religions operate as social-ideological systems with established beliefs, practices, roles, and institutions that are densely interconnected, mutually reinforcing, developed over time, and embodied in material culture. The systemic, dynamic, interactive features of religion are thereby lost to view and so is the significance of those features in the ongoing trajectories of religion in the contemporary world—a world in which large-scale economic and demographic shifts and other supra-individual phenomena and dynamics are crucial, not to say cataclysmic. What is likely to prove especially significant in that regard is the interplay between such large-scale shifts and the force of particular ethnic, national, and ideological allegiances, including emergent social and political alliances and new forms of social identity and sociality. In seeking to grasp the effect of such multi-scaled dynamics on what we call "religion" or the place of specifically religious beliefs in this complex interplay, we require more extensive and diverse intellectual resources—empirical, theoretical, analytic, and interpretive—than those currently offered by "cognitive" or "evolutionary" approaches alone.

A second general point may be stressed here. In conducting his bio-economic analysis of religion, Atran's human universals are more than a little over-generalized. Thus, although he assesses religious beliefs and practices in relation to what he sees as "the moral and existential dilemmas that panhuman emotions and cognitions force on human awareness and social life" and invokes the "moral sentiments and existential anxieties" that "constitute—by virtue of evolution—ineluctable elements of the human condition," he fails to register what are likely to be the central existential concerns and related sentiments of his own readership—that is, contemporary, educated, more or less secular and more or less reflective Westerners. I mentioned above Atran's failure to acknowledge or account for the ethical behavior of nonbelievers, that is, those who, in apparent defiance of his analysis, behave honestly and otherwise honorably even though they don't believe "the gods are watching." Of course, nonbelievers know that other *people* are watching, but that does not explain their more than occasional fashioning of strenuous ethical principles, including principles, such as personal, ideological, or political loyalties, that are known only to the persons who live by them. Also, although Atran writes of the existential fears allayed by religious concepts and the self-interested impulses controlled by religiously sanctioned social norms, he says nothing about the significant desires and impulses that are thereby thwarted—desires, for example, for individual autonomy, non-normative social roles, or access to proscribed knowledge. To be sure, such desires may not be "panhuman." Nevertheless, they may figure quite centrally among the existential conditions of the humans who experience them and have certainly figured crucially in the social and political dynamics of post-Paleolithic human history. Accordingly, such desires, along with other culturally, socially, and historically significant—though not necessarily universal—needs, sentiments, principles, and behaviors, must be included in tallying the costs and benefits of religion, whether to predict the likely survival or disappearance of the gods or to engage other intellectual concerns.

"THE GODS SEEM HERE TO STAY"

THE FUTURE OF "RELIGION"

It is normal for people to believe in the works of the imagination. People believe in religions, in Madame Bovary *while they read, in Einstein, in Fustel de Coulanges, in the Trojan origin of the Franks. However, in certain societies, some of these works are deemed fictions. The realm of the imaginary is not limited to these. . . . Daily life itself, far from being rooted in immediacy, is the crossroads of the imagination.*

—*Paul Veyne,* Did the Greeks Believe in Their Myths?

Given the general complexity and fundamental heterogeneity of what we call "belief," it is by no means clear how to identify a specifically *religious* belief. Such identification becomes even more elusive in view of the multiple, continuously—and, now, exceedingly rapidly—changing forms and manifestations of what we are prepared to call "religion." Given both problems, it is hard to say what would or should count as the survival or disappearance of religious belief in human history. To approach these questions, we may return to two related issues noted earlier. One is the New Naturalist identification of religion with concepts of supernatural agency. The other is the assumed transparency and transhistorical stability of the term "supernatural."

In company with Boyer, Dennett, and other New Naturalists, Atran maintains that, as he writes, "Supernatural agency is the most culturally recurrent, cognitively relevant, and evolutionarily compelling concept in religion." His version of their shared explanation of that concept is worth reviewing:

> The concept of the supernatural agent is culturally derived from innate cognitive schema, "mental modules," for the recognition

> and interpretation of agents, such as people and animals. . . . A naturally selected, mental module is functionally specialized to process, as input, a specific domain of recurrent stimuli in the world that was particularly relevant to hominid survival. . . . It allowed humans to adaptively navigate ancestral environments by responding rapidly and economically to important, statistically repetitive task demands, such as distinguishing predator from prey and friend from foe.[38]

Even apart from its familiar invocation of dubious input-output mental machinery, this explanation raises important questions of conceptualization. There are, of course, many sorts of non-natural, super-ordinary, or other-worldly beings—plus forces, realms, and states—that figure in human experience and cultural lore (for example, angels, grace, Nirvana, Hades, Fairyland, and the Force), and only some of them are "religious" in any standard sense of the term. Moreover, only some of these entities (or "ideas," "concepts," or "constructs") are person-like or animal-like agents and, among such agents, along with sun-gods, earth-spirits, Jove, and Jehovah, are mermaids, giants, sphinxes, and extraterrestrials. The standard New Naturalist account implies, and Atran and Boyer might readily affirm, that all these anthropomorphic or zoomorphic agents are, in Atran's phrase, "culturally derived" from a single set of innate mental modules or cognitive mechanisms. But, if this is so, then clearly the category of *supernatural* agents thus derived cannot be identified with what either of them (or virtually anyone else) otherwise refers to as "religion."

We may also review here Boyer's summary of his evolutionary-psychological explanation of religious concepts, as quoted in chapter 2 above: "Some concepts happen to connect with inference systems in the brain in a way that makes recall and communication very easy. . . . Some concepts happen to connect to our social mind. Some of them are represented in such a way that they soon become plausible and direct behavior. The ones that do *all* this are the religious ones we actually observe in human societies."[39] The point to be noted here is the disparity between the scope of Boyer's descriptions and the validity of his identification or quasi-definition.[40] For, if the inference systems and mental dispositions thus described do exist, there is no reason to think that the concepts that issue from their joint operation ("the ones that do *all* this") are specifically "the religious ones." Rather, the operation of such mechanisms in various combinations and configurations would appear to be reflected in concept formation more generally, including the formation of such familiar secular concepts as Lady Luck or Spider-Man[41]—and, indeed, in the formation of such sophisticated concepts as invisible hands, selfish genes, and the New Naturalists' own Natural Selection represented as a supremely powerful quasi-purposive quasi-agent.

Related questions about New Naturalists' accounts of the supernatural are raised by ethnographic observations that challenge their shared conceptualization of the counterintuitive, including the identification—explicit in Boyer and Atran—of supernatural entities with a conjunction of intuitive and counterintuitive properties. For example, anthropologist Maurice Bloch notes that, among many people who worship supposedly supernatural ancestors, attitudes toward such ancestors are not very different from attitudes and behavior toward living elders: "The motivations, emotions, and understanding of elders and ancestors are assumed to be the same. Ancestors are simply more difficult to communicate with. Thus, when rural Malagasy, in perfectly ordinary contexts, want to be overheard by the dead, they speak more loudly, something they often also do when they want elders to take notice, since these are also often deaf. . . . The ancestors are not as close as living parents or grandparents, but they are not all that distant."[42] Bloch observes more generally that properties of gods, ghosts, or spirits that are acknowledged by people (including Europeans) in ritual or mythological contexts are often doubted or rejected by them in other contexts (for example, academic or professional), and that ideas that were at one time strange, striking, and thus counterintuitive for someone (for example, the idea of a personal deity who is invisible, sees everything at once, rises from the dead, and lives forever in a world of spirits) may become, for the born-again or otherwise converted person, familiar, routine, and thus intuitive.

We may also note that, contrary to a standard New Naturalist account of the operation of intuitions, not all cognitive norms—that is, the "default settings" in relation to which we experience concepts as either intuitive or counterintuitive—are innate, universal, or hardwired in the human brain. Many such norms, for example those involving gender-appropriate behavior, are clearly post-natally acquired and significantly shaped by individual experiences in particular social and cultural environments: "How awful that their women have to cover themselves up like that!" or "How terrible! Those women are not properly veiled." Moreover, all cognitive norms, including those with arguably genetically specified aspects, can be significantly adjusted over the course of a lifetime. Sometimes, as in the case of gender-appropriate behavior, this can occur as a result of expanded experience in different social environments. Or, as in this and other cases, an adjustment in the default setting of a cognitive norm can occur in responsive correspondence with other norms subsequently learned or otherwise developed, for example, ethical norms of tolerance for physical or social difference.

Contrary, then, to the assumption or claim—articulated by both Boyer and Atran—that the intuitiveness or counterintuitiveness of a concept such as god, elf, burning bush or plumed serpent is the automatic outcome of the activity of a universal cognitive mechanism responding to the inherent properties of some domain of stimuli, it seems that our experience of any concept as either (or more or less) intuitive or counterintuitive—as cognitively consonant or cognitively dissonant, as more or less proper or disturbing—is, to a significant extent, culturally contingent and individually, contextually, and developmentally variable. The emphatically reiterated claim by Atran, Boyer, Tooby, Cosmides, Pinker, and other evolutionary psychologists that these and all other cultural, historical, contextual, and individual multiplicities, differences, contingencies, and variables are rendered negligible or strongly "constrained" by the cross-cultural, transhistorical, uniform operation of evolved, innate, universal cognitive mechanisms seems to be just that: a claim, an assumption, a conviction, a dogma.

Finally, here, we may note the difficulties involved in the New Naturalist identification of supernatural agents with a *conjunction* of intuitive and counterintuitive properties. Explaining the point, Atran writes: "Supernatural agents are always *humanlike* but never quite human."[43] Certainly one finds throughout history presumptively mortal beings with the standard default properties of persons who nevertheless appear also to have been endowed with extraordinary and not-quite-human characteristics: Shakespeare and Mozart, Jesus and Gandhi, Napoleon and Mao, Garbo and Callas—geniuses, prophets, superheroes, and superstars fixed in the cultural-cognitive heavens, their powers the object of enduring awe and sometimes enduring worship. As Bloch puts it, "the counter-intuitive . . . is everywhere."[44] The same may be said, for comparable reasons, for the supernatural.

But if, as seems to be the case, the counterintuitive and the supernatural are everywhere, then where in particular, we may ask, is the New Naturalists' "religion"? The answer, I would suggest, is that it is nowhere in particular because everything that is said to define or comprise it is found everywhere in human experience and culture. If that is so, then clearly the gods are indeed here to stay, for their existence and survival appear coextensive with our own. This is not, however, because we are hardwired to generate and worship deities or because humanity depends for its survival on a general belief in a traditionally conceived supreme being. It is, rather, because there is nothing that distinguishes how we produce and respond to gods from how we produce and respond to a wide variety of other social-cognitive constructs ubiquitous in human culture and central to human experience.

CHAPTER FOUR

Deep Reading: The New Natural Theology

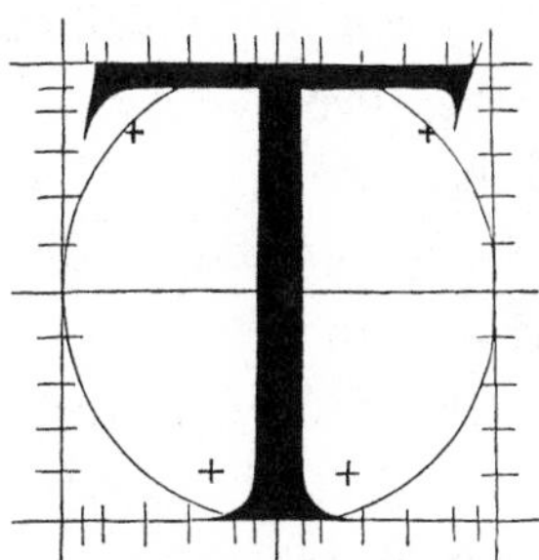

This chapter has a double focus: two intellectual projects, each involving the relations between theology and contemporary science, one essentially reconciliatory, the other distinctly antagonistic. The first is a set of efforts, primarily by scientifically knowledgeable theologians but also by some theologically inclined scientists, to reveal a cognitively satisfying consonance between the accounts of nature given in the natural sciences and traditional Christian belief. Examples of such efforts include *Exploring Reality: The Intertwining of Science and Religion* by quantum physicist/Anglican priest John Polkinghorne; *Theology for a Scientific Age* by bio-chemist/Anglican priest Arthur Peacocke; and, of particular interest here, a widely cited book by the Catholic theologian John Haught titled *Deeper than Darwin: The Prospect for Religion in the Age of Evolution.*[1] The second focus is a range of rebarbative writings—expressions of irritation, formal critiques, and attempted rebuttals—by theologians and lay theists responding to current naturalistic, especially Darwinian, accounts of religion. As will be seen, there are a number of points of interest here for the general concerns of this book, including contemporary views of cognition and belief. I frame them below largely through an examination of Haught's *Deeper than Darwin*, taking it first as an example of current theological readings of natural science and then as representing critical responses by theistic intellectuals to the New Naturalism.

READING EVOLUTION

Attempts, as it is said, to "reconcile" science and religion have a history in the West as long as the independent existence of each.[2] Given that history and the shifting meanings of several key terms involved, I should note some important differences between what I am calling here the *New* Natural Theology—understood as contemporary efforts of that kind—and older, more familiar versions of natural theology as such. Traditional natural theology—articulated by Augustine, pursued by Thomas Aquinas, elaborated most influentially under that name by William Paley in the early nineteenth century, and rehearsed in our own time by proponents of so-called Intelligent Design—seeks and claims to find evidence in nature alone, as distinct from scripture or revelation, for the existence and attributes of a supreme being. In contrast, the aims and claims of the New Natural Theology are subtler and more modest but also, by the same token, more equivocal and elusive. Thus Haught and Polkinghorne accept without reservation the accounts of nature framed in contemporary natural science but "scrutinize" them (the word is Haught's) for theological import, seeking not evidence of God's existence but, in a phrase they share, merely "room" for a providential deity; not proof of redemption but possibly legible "signs" of a "surprising" future; not grounds for believing in personal immortality but "reason" to "hope" for the fulfillment of what we "intuit" as a certain "promise."[3] Here as elsewhere in the New Natural Theology, scholastic logic and traditional hermeneutics are supplemented by more contemporary modes of deep reading while confident displays of theistic triumphalism are replaced by the ambiguous gestures of theological-scientific compatibilism.[4]

The New Natural Theology is largely a rhetorical enterprise, a matter of making a series of complex, somewhat paradoxical ideas credible through a skillful use of language. This often involves the construction of conceptual syntheses through an interweaving of idioms. Thus Haught, like Polkinghorne, Peacocke, and other scientist-theologians working in the genre, deploys the technical terms of contemporary natural science either in smooth conjunction with elements of a more familiar and resonant religious idiom—scriptural, liturgical, or homiletic—or with an expressly theological spin. In a typical formulation, Haught writes: "The divine fidelity to promise is always and everywhere intimately embedded in and creatively working through the laws of nature as well as the novelty that breaks in through the accidental." Similarly, joining terms from linguistics to distinctively Christian concepts, he writes: "Nature's predictable laws . . . can . . . be read as necessary grammatical rules that any incarnation of deeper meaning must adhere to if it is to receive embodiment." A bit later, invoking the technical term "deep time" from geology, he observes: "From a theological perspective, deep time may be thought of as the consequence of God's opening up the world to a luxuriantly creative future."[5] In these examples and more generally, the resulting turns of phrase may be evocative for many readers but are likely to be conceptually thorny and, especially for skeptical readers familiar with scientific ideas, to sound more or less peculiar.[6]

A second significant feature of the New Natural Theology, though itself far from new, is the practice of allegorical explication. A central resource of traditional hermetic, scriptural, and literary hermeneutics, it reflects the idea that a particular text—book of Nature, God, or mortal author—delivers its messages "otherwise" and thus requires oblique and/or expansive reading. It operates, in other words, on the assumption that what something means most importantly is different from what it means when read literally, conventionally, or, in Haught's preferred term, "shallowly." In the New Natural Theology, allegorical explication can be pursued at many levels and in several directions, though all ultimately convergent. Thus scriptural passages (notably Genesis 1–9) are interpreted so as to resonate with current understandings in cosmology, geology, and paleoanthropology while such natural phenomena as the Big Bang, the appearance of living forms, or human evolution are read both as divinely ordained events or processes and also as divinely scripted messages to humanity. Where nature and scripture are understood as having the same author, they can be readily subjected to mutual and mutually reinforcing exegesis.

A third feature of current efforts to negotiate between scientific knowledge and religious doctrine is the recurrent observation that "room is left" for theistic postulates in the accounts of nature offered by contemporary science: for example, in cosmology, quantum physics, and chaos theory as interpreted by Polkinghorne and in neo-Darwinian evolutionary theory as read by Haught. Thus Haught writes:

> Even after we have assented to the impressive theory of evolution by natural selection . . . it still leaves many questions unanswered. And while we can respond to some of these by moving into other natural sciences . . . , even a thicker scientific account will still leave abundant room for a theological reading of life as well.
>
> . . . In the first place, biological evolution requires a universe that is open to accidental, undirected or "contingent" events. For example, nature has to allow room for the novel combinations of chemicals that permit the spontaneous origin of life. Then, after life appears, there must be room for random mutations in genes.[7]

Two different ideas should be distinguished here. One, familiar from arguments for Intelligent Design, is that wherever a phenomenon is found that cannot be given a generally accepted naturalistic explanation, an essential inexplicability may be inferred and the workings of something beyond nature supposed or affirmed—a chain of reasoning that yields what is called, in more skeptical circles, "the God of the gaps." The second, subtler idea, characteristic of the New Natural Theology, is that, when read appropriately, key explanatory concepts in current scientific accounts of nature can be recognized as already theologically fraught. Thus, for Haught, scientific understandings to the effect that the origin of life was "spontaneous," that the combination of chemicals that gave rise to life was "novel," and that the mutations responsible for genetic variation are "random" can be read individually and together as suggesting that certain events and processes are outside the laws of nature (taken by Haught as entailing strict regularity and predictability or, as he puts it here, "draconian determinism") and, accordingly, that there are non-natural agencies, operations, or states for which those laws must "leave abundant room."

The notion of room left (or not) for theistic postulates in scientific descriptions of nature suggests a more general conception of a finite space, presumably logical or ontic, where different accounts of the world—here, naturalistic and theological—jostle one another for limited quarters. This spatial conception of the relation among alternative ontologies is recurrent and ubiquitous. It is found both among theologians, who may lay claim to such room on the ground that a strictly naturalistic description of the world does not amount to the whole truth, and among metaphysical naturalists, who may deny such room on the ground that only one description of the world can be true. The question may be raised, however, as to whether that shared topographical conception does not foster, both for the reconciling theologian and for the exclusivist naturalist, the very intellectual perplexities that its invocation is supposed to resolve. I return to the question below. First, however, we may look more closely at Haught's expansive reading of contemporary evolutionary theory.

In a central section of *Deeper than Darwin*, Haught develops the idea that biological evolution constitutes a narrative or story. The unfolding of this narrative, he observes, requires the prior assembly of three crucial ingredients, namely, "contingency," "invariance," and "deep time," while the plot unfolds in such a way that the story's protagonist, life, undergoes a series of transformations over a long period of time, ultimately manifesting itself as humanity. In the very manner of its unfolding, he continues, "the story of life" can be seen to foretell that, after another perhaps very long period of time and undergoing further transformations, we (humanity as such or perhaps just some of us—the referent is unclear) will arrive at the destiny promised to us—presumably by scripture and Christian tradition, though the source of the promise and the destiny itself remain unspecified. Among other notable features of this reading of evolutionary theory is Haught's repeated use of the term "unfold" to describe biological events and processes. As distinct from the historian's "occur," the storyteller's "happen," or the biologist's "develop," the term carries strong teleological echoes. To the historian of science, it might suggest preformationism, the view that an organism's mature features are folded up into and somehow fully predetermined by its initial form. To the literary scholar, it might suggest a duly informed explication (unfolding or folding-out) of an author's intended meanings. To the theologically instructed reader, however, the term is most likely to suggest the manifestation in historical time of the divinely and eternally ordained.

Rehearsing and commenting on his reading of the Darwinian account of evolution, Haught writes:

> [N]ature has revealed itself . . . as being an immense story. . . . And this story, in turn, may very well be open to many levels of reading.
>
> To be perfectly clear here, Darwinism *presupposes*—since by itself it cannot account for—the narrative cosmic tablet on which the life story becomes inscribed. Or, to put it in other terms, our three background ingredients [that is, "contingency," "invariance," and "deep time"] must be waiting on the cosmic table long before Darwinian process begins mixing and cooking them. . . . Neither Darwinian science nor any of the other sciences can satisfactorily tell us "why" nature is constituted in just such a way

> as to allow the universe to unfold narratively . . . [T]he question . . . invites theological comment.
>
> . . . [It] could then be thought of . . . as purely accidental. . . . [But the question is] why any story at all? The point is, we cannot help embedding all of our speculation . . . within a "narrative *a priori*" in accordance with which evolution has forged our own brains.[8]

Two rhetorical features of these comments, along with their combined effect, are worth noting. One is the joining of expressions of intellectual tentativeness (such as "this story . . . may well be open to many levels of reading" or "[It] could then be thought of . . . as purely accidental") with what are represented as logical, ontological, or psychological necessities ("Darwinism *presupposes*," "must be waiting on the cosmic table," "we cannot help embedding"). Another is the weaving together of the relatively neutral, descriptive language of science ("evolution," "Darwinian process," "nature is constituted," and so forth) and the heightened, metaphoric, oblique, and archaic language of religious doctrine ("Nature has revealed itself," "the cosmic tablet," "the life story becomes inscribed," and so forth). Taken jointly, these features give a sense of combined scientific weight, objective necessity, and poetic resonance to an account that might otherwise be seen as a series of gratuitous inferences tenuously strung together.

At the conclusion of his reading of the evolutionary narrative, Haught suggests that the author of the story of life arranged for our brains to be constructed so that we, or at least those of us appropriately prepared, cannot help but recognize (his term is "make out") that story when we study nature: "When we read the universe with eyes and hearts prepared by the motif of promise, there emerges a palpable consonance between the narrative character of a life-bearing evolutionary cosmos, on the one hand, and the general thrust of religious hope on the other."[9] Or, it could be said, when we read the universe with perceptions and emotions primed by stories of providence and redemption, then the universe tells us stories of providence and redemption. Haught evidently means to suggest that this consonance is a sign of God's hand in the creation of both the universe and the brains of the creatures—ourselves—who can read it. His comment here, however, can also be seen as an implicit acknowledgment of the circularity at the heart of the new natural-theological hermeneutics: one finds what one already has; what is revealed is what one already believes.

Haught's deep reading of biological evolution will be cognitively agreeable to those prepared to accept it. These are largely people for whom the major elements and characteristic idioms of Christian belief and practice are intimately familiar and who operate with those elements and idioms every day—conceptually, perceptually, emotionally, pragmatically, socially, and, of course, verbally. Conversely, to those who come to such readings not so primed, who operate every day in all these ways with different beliefs and idioms, *Deeper than Darwin* and other works of New Natural Theology will sound not—or not necessarily—wrong or foolish but, to repeat a term I use above, *gratuitous:* sources not of ultimate meanings agreeably revealed but of possible interpretations ingeniously constructed. Nothing in nature or logic can prevent anyone from giving theological glosses to the findings and concepts of natural science. But, equally, nothing in logic or nature requires anyone to accept them. The theological deep readings—allegories, divinations, ways of seeing the universe, ways of putting together convictions and forms of knowledge—remain optional, the products of imaginative possibility, not of any presumptive necessity.

In view of the apparent inevitability of these diverse modes of reception, Haught's theological gloss of evolutionary theory and similar current efforts to reconcile science and religion cannot demand or expect widespread, much less universal, uptake. And, in fact, neither Haught nor the New Natural Theologians more generally make any such demands or seem to have any such expectations for their works. Rather, as appears to be tacitly understood by all concerned, these elaborate cognitive conceits are highly particular in destination: offered by those who can construct them to those who can use them. It would follow that different deep (or "shallow") readings of the universe—for example, pantheistic, polytheistic, or strictly naturalistic—might prove no less cognitively consonant for those whose eyes and hearts are otherwise prepared. Or, we could say, there is always "room left" for alternative ontologies in cognitive-intellectual space, a realm that is neither cramped nor finite but, on the contrary, appears—both historically and for humans individually—exceedingly and perhaps infinitely elastic. The theological/metaphysical-naturalist view is that room is left (or not) for alternative constructions in nature, reality, or the universe. The view I am suggesting here is that all of these, not only the rival ontologies but also the roomy or confined ontic spaces they appear to occupy, are products of our own cognitive-intellectual-imaginative activities.

The effort to reconcile science and religion responds to the claim or fear that, under the assumed conception of each, they are in unhappy conflict. It is not obvious, however, that such a conflict exists, at least not in the ongoing lives and experiences of individuals as distinct from the logical or ontic spaces posited by theologians or metaphysical naturalists. Difficulties arise under two very particular conjoined conditions: first, when "science" and "religion" are understood as consisting, respectively, of duly established scientific accounts of the natural world and duly established sacred doctrine; and, second, when it is assumed that spiritual or mental hygiene requires that all our ideas, impulses, affections, and acts be mutually aligned all the time. For many people, however, accepting, applying, and/or producing scientific knowledge and being religiously observant are no more in conflict than would be, for any of us, both playing the violin and practicing law. They are, rather, two different kinds of things that one may do and/or be: activities performed, identities played out, and experiences sustained in different contexts, each involving different cognitive and bodily configurations, corresponding to different capacities and desires, and offering different forms of satisfaction. So understood, a conjunction of science and religion—engaging in some significant way with scientific knowledge, accepted as such, and also in some significant way "being religious"—need not involve acute epistemological difficulties or cognitive dissonances or, accordingly, require the reconciling of anything with anything else. While all formal reconciliations of "science" and "religion" appear to be conceptually fragile and certainly require heavy rhetorical scaffolding to hold them in place, the ongoing inhabiting of both is, for many evidently well-functioning people—scientists and science teachers, anthropologists and historians, doctors and engineers—a daily fact of life.[10]

THEOLOGY AND THE NEW NATURALISM

Scholars in the field of religious studies have greeted the new cognitive-evolutionary approaches to religion with a variety of responses, ranging from swift dismissal to cautious interest or outright enthusiasm. Theologians, at least those who have taken note of these approaches, have tended to regard them with acute distaste and, in a few cases, have sought to subject representative works to formal critique and rebuttal. Objections are commonly raised to what are seen as the evolutionists' biological reductionism, spiritual shallowness, and ignorance or ignoring of religious people's experience, while the rebuttals commonly consist of efforts at metaphysical outflanking or demonstrations of alleged fatal self-contradiction.[11] Haught's responses in *Deeper than Darwin* to what he calls the "new Darwinian" explanations of religion are in many respects typical along just these lines. They are of particular interest here, however, because Haught has made evolutionary theory the focus of his own efforts as a theologian and because his responses highlight some significant intellectual liabilities of the evolutionary approaches themselves.

In a central chapter of *Deeper than Darwin*, Haught attempts to undercut the claims of evolutionists to explain various aspects of religion. The term "undercut" is especially apt here since a good part of the issue is epistemic rank as marked by comparative depth: in effect, the lower the depth, the higher the rank; the deeper, the truer. Thus Haught titles a section of his critique "Deeper than Deep Darwinism." Not irrelevantly, however, a number of evolutionary psychologists participate in this competitive depth-sounding. For example, John Tooby and Leda Cosmides, in an inaugural essay titled "The Psychological Foundations of Culture," claim underneath-it-all status for evolved mental mechanisms and, with it, methodological priority for their adapted-mind program of evolutionary psychology. "Only after the description of the evolved human psychological architecture has been restored as the centerpiece of social theory," they write, "can the secondary . . . effects of . . . social dynamics be . . . analyzed."[12] We may also recall here the "deep structures" of Chomskian language theory, adopted by Lawson and McCauley to explain religious rituals, or, in Boyer's *Religion Explained*, the "mental basements" where all the "underlying" causal mechanisms are said to be located. These evolutionary theorists evidently share with Haught a classic surface-and-depth model of phenomenal appearance and ontological reality, though of course they identify that underlying reality in quite different ways.[13]

It is not surprising that a Catholic theologian would find strictly naturalistic accounts of religious beliefs and practices—his own among others—cognitively dissonant and, when coupled with bumptious titles like *Religion Explained* or *How Religion Works*,[14] disagreeable in the extreme. What is of interest here is Haught's effort to translate that reaction into a dispassionate intellectual critique. One sign of the strain of the effort is his sardonic tone, which is otherwise uncharacteristic but becomes, in this chapter, a persistent and evidently irrepressible note of sarcasm. Another is his repeated mischaracterization of the evolutionists' aims and claims.

Introducing his discussion of the new evolutionary accounts, Haught writes: "[W]e can now explain 'the persistence of gods' . . . without assuming the hidden influence on human consciousness of any ontologically real divine presence." To a theistic fundamentalist, explaining *anything* without assuming the influence of a real divine presence would figure as impiety, and the exclusion of such an assumption would constitute an objectionable feature of any strictly naturalistic teaching as such. Haught is not a theistic fundamentalist. He is a scientifically informed theologian seeking explicitly to reconcile Darwinian accounts of the world with traditional religious teachings. His project pulls up short, however, at Darwinian accounts of religion itself. He continues: "All of the characteristics we associate with religion—its rituals, doctrines, stories, institutions and theologies—are fully understandable, then, if we read them as having had an adaptive function."[15] This is a significant overstatement. As we have seen, what are said to have had an adaptive function are not the "rituals, doctrines, stories, institutions [or] theologies" themselves but, rather, a number of more general, presumably innate and universal human traits—intuitions, cognitive mechanisms, behavioral dispositions, and so forth—that are posited as having conferred fitness advantages on our hominid ancestors.[16] The central claim of the Darwinians cited by Haught is that we can understand the recurrence and persistence of many features of religious belief and practice if we see them as products or byproducts of the continuing operation of such otherwise adaptive mechanisms and dispositions. Many aspects of this claim, and of the New Naturalist project more generally, can be challenged on conceptual, methodological, or empirical grounds. But Haught does not mount any such challenge here. Rather, like a number of other theological and lay theistic commentators, he attempts to make Darwinian accounts of religion look absurd and reductive by exaggerating the scope of their claims ("all," "fully") and by oversimplifying their central points. Ultimately, as I discuss below, he attempts, employing a recurrent tactic, to discredit such accounts as self-refuting.

Haught's descriptions of current evolutionary explanations of religion are, as just indicated, in many respects inaccurate. Significantly here, however, some of his mischaracterizations are fostered by dubious moves and careless formulations in the evolutionists' own accounts. One such move is the wholesale adoption of the two-story (upstairs/downstairs, surface/depth) model of human behavior and culture familiar in evolutionary psychology. Thus, in Boyer's *Religion Explained* and elsewhere, phenomena ranging from the performance of communal rituals, including the Catholic Mass, to the experience and display of social emotions, including affection for fellow members of a religious collective, are represented as superficial manifestations of the operation of automatic and unconscious as well as primitive psychobiological mechanisms. For Haught, such accounts are clearly offensive. One needs no theistic commitments, however, to find them explanatorily thin.

Another feature of the evolutionists' accounts of religion that commonly invites misunderstanding is the metaphoric representation of gene-level selection as the purposeful action of ingenious genes. Haught writes: "Darwinian analysis allows anthropology to conclude that religion, beneath its complex surface manifestations, was *ultimately* invented by our genes as an adaptive contrivance. . . . In Burkert's understanding, once again, the gods persist because of our genes' need to persist."[17] No Darwinian would maintain that religious belief arises or persists because of some literal strategy on the part of human genes. Indeed, a number of evolutionists have stressed that there are no "genes for God" or genes "for" any other specific aspect of religion as such.[18] Nevertheless, Richard Dawkins's notion of "selfish" genes is casually invoked by Burkert in *Creation of the Sacred* and by other evolutionists in connection with their explanations of religion and other forms of human behavior. The unhappy effect of the all-too-vivid metaphor on Haught and in the popular imaginary is considerable.

In a celebrated passage at the end of *The Origin of Species*, Darwin observes a "grandeur" in the de-theologized, de-teleologized account of the development of living forms he has just laid out. It is a kind of grandeur—epic or indeed biblical—that Haught attempts to regain for Christians by re-theologizing the Darwinian account, giving it an explicitly teleological narrative structure and infusing it with Christian concepts and idioms. One can understand his distress, then, at the insistently "tough-minded," "hard-nosed" retelling of that evolutionary tale by Dawkins, this time with genes, not a benign Mother Nature or even a mighty Natural Selection, as the chief agents. To an imaginative theologian like Haught, it would sound like a particularly nasty version of the *Götterdämmerung*, with control over living forms transferred from an exalted, providential deity to a gang of greedy little Nibelungen working busily underground—or, indeed, and more disagreeably, deep within each of us—to further their own purposes.

The picture of the universe that emerges from what Dawkins and others after him call "Darwinian thinking" is not merely of no God, no soul, and nature red in tooth and claw, but of every feature of every creature—bird, tree, squirrel, and strawberry, one's child, one's lover, and one's self—as being, top to toe, inside and out, the way it is only because better enabling the survival and reproduction of the genes that specify it or, at best, because carried along by such a functionality.[19] The image of life thus offered is, to say the least, disconcerting; and one may sense a host of cognitive mechanisms rushing in to supply other images or alternative conceptualizations that provide an understanding of the living world—including one's own self—more in accord with familiar orientations. Those alternative images or ideas need not, however, be fictitious or false by naturalistic standards. Thus we may remind ourselves that, although our general structures and modes of operation as biological creatures have been strongly shaped by selection pressures, not everything we do as particular persons involves the furthering of our own reproductive fitness or the perpetuation of our genes. We may also remind ourselves that, as creatures who continue to develop throughout our lives, we are affected by particular experiences that shape our responses, purposes, judgments, and actions—and, indeed, our bodies, top to toe, inside and out—no less significantly than do our biological endowments. Although a rigorous pursuit of so-called Darwinian thinking can be cognitively disturbing, a duly rigorous restorative activity does not require a rejection of either the logic of Darwinian theory or its implications for the phenomenal world.[20] Conversely, and contrary to the apparent supposition (and strenuous example) of some Darwinians, nothing, including a due appreciation of that logic and those implications, obliges anyone to engage in such thinking all the time.

In another aggrieved passage related to the one cited above ("once again, the gods persist because our genes need to persist"), Haught writes: "A usually unspoken but clearly operative postulate in this new Darwinian interpretation [of religion] is a quite modern one: . . . religions provided illusory but effective shields against the terrors of existence. And by favoring our species with the fictitious phantasm of a purposeful universe, religions gave our human predecessors a reason to keep on living, to bear offspring and thus keep their genes from perishing."[21] Of course, the idea that religions provided our ancestors with shields against the terrors of existence is not modern: it figures in virtually every naturalistic account of religion since Lucretius. Haught gives no citation for the assumption he frames here, but is likely thinking of Boyer's general adaptationist account in *Religion Explained* combined with a number of passages in Burkert's *Creation of the Sacred*.[22] Burkert speaks rarely of genes and does not brush off ancient myths as "fictitious phantasm[s]," but he does suggest that a major source of the power of religions, modern as well as ancient, is the sense of order, stability, and manageability they give to our experience of an otherwise chaotic, inexplicable, and often catastrophic world. Thus he writes in a characteristic passage: "[W]e are unavoidably dependent upon a variety of circumstances both known and unknown. . . . Religion makes all this secondary by turning the attention structure toward one basic authority . . . [thus] creating sense out of chaos." Similarly, in a later passage, Burkert writes: "Modern science . . . will not easily prevail. People prefer to cling to the surplus of causality and sense." And, at another point, discussing gift exchange, he comments: "The principle of reciprocity . . . is not only a . . . widely successful [social] strategy but a postulate acted out to create a stable, sensible, and acceptable world, gratifying both intellectually and morally and bridging the gap of annihilation. . . . Thus teaching and rehearsing . . . [such] strategies has widely prevailed in religious tradition.

Life is bound to optimism—even this may be called a biological necessity."[23] If, however, these are the sorts of comments that Haught has in mind, then his paraphrase of them in the passage quoted above transforms a series of fairly urbane reflections on the complex origins and operations of religion into raw biological reductionism.

Toward the end of his critique of current Darwinian accounts of religion, Haught produces a barbed contrast between the authors of such accounts and earlier critics of religion:

> [Our religious ideas] are pure fictions, and lovers of truth should be willing to give them up—as apparently the new Darwinian debunkers of religion have done themselves.
>
> However . . . [w]hile they claim to be in possession of the final truth about religion, they are not terribly disturbed, as were earlier critics of religion, that most of humanity still wallows in the essentially false comforts of religion.

Those earlier critics, he adds, would have found this new view—that is, what Haught represents as the Darwinians' blasé attitude toward what they must regard as the benighted state of most of humanity—"highly objectionable on moral grounds." Haught is, of course, being doubly derisive here, suggesting that the earlier critics of religion, wholly objectionable though their views and explanations were, had a certain moral edge over contemporary Darwinians. For the old debunkers (presumably Marx and Freud among others) at least put an old-fashioned high value on truth, while the new ones, though evidently convinced that religious ideas are illusions, seem not to care that so many people still "wallow" in them.[24]

The situation here is complex. First, it is true that a

number of Darwinians, including Boyer and Burkert, exhibit what appears to be a biologically instructed conviction of the wholly naturalistic springs of religious devotion combined with an anthropologically or historically instructed conviction of the value of such devotion for humans generally. There is, of course, nothing self-contradictory in that conjunction (as Haught seems to want to suggest) or surprising in the fact that those so persuaded are disinclined to aggressive debunking. It does raise the question, however, of the Darwinians' ethically proper intellectual relation to most of their fellow mortals. Thus it could be maintained that, as scholars or scientists, they are obliged to be truth-tellers, to give their fellow mortals the benefit of their discoveries about the primitive psychobiological springs of religious ideas and practices. But it could also be maintained that, as scholars and scientists, they are obliged to act in accord with the knowledge they have of the positive value of religious ideas and practices, especially since it is a value that they have good reason to think is not only individual and subjective but communal and substantive in its effects. This latter sort of empirically—historically, ethnographically—grounded recognition of the value of (other people's) religious convictions is not the same as the forms of ethically or epistemologically grounded civility or tolerance often associated with "liberal pluralism," "multiculturalism," or "cultural relativism." But it is not altogether different from them either.

What Haught represents as the new Darwinians' nonchalant attitude toward truth seems to be just the fact that a number of scholars and social scientists pursuing evolutionary approaches to religion do *not* frame their accounts as Enlightenment revelations or use them as platforms to denounce religion generally. There is an important corollary here that escapes Haught's notice, though it has been recognized and welcomed by various theistic intellectuals (sometimes under the label "postmodernism") and, for the same reasons (and often under the same label), reviled by various scientistic naturalists and polemical atheists.[25] For, precisely to the extent that scholars studying religion reject the absolutist, exclusivist, self-privileging conceptions of truth shared by theologians like Haught and scientists like Dawkins, the more likely they are to register both the pragmatic, aesthetic, and/or epistemic value of various religious ideas and practices and also the forms of human ingenuity, artistry, and imagination involved in their past and ongoing elaboration. It appears to be this appreciation of various aspects of religion by academics or intellectuals who are not necessarily themselves observant that, among other things, Daniel Dennett refers to as "belief in belief" and condemns as faint-hearted complicity with the forces of superstition and unreason.[26]

Toward the conclusion of his critique of evolutionary accounts of religion, Haught produces against their authors the customary—purportedly *coup de grace*—charge of self-refutation. Thus he asks pointedly, "Can the evolutionary psychologist coherently naturalize human culture, which includes science, without sabotaging the authority of Darwinism itself?"[27] Here, as often elsewhere, the charge is enabled by an absence of explicit reflexivity on the part of those offering otherwise challenging views. In this case, it is enabled by the failure of those offering evolutionary explanations of religious beliefs to consider the implications of such explanations for their own beliefs as scientists. That absence or failure of reflexivity, especially where it is attended by implicit or explicit self-exemption, marks a weak spot in the New Naturalist accounts of religion. But it does not constitute a decisive self-undermining of those accounts in the way that Haught and various other theistic critics seek to demonstrate.

The premise of Haught's argument is that the epistemic authority of Darwinism, its claim to offer compelling scientific knowledge of, among other things, religion, must have a transcendent warrant: something like special access to underlying truth or a method for discovering objective reality; some grounding that exceeds the merely natural, merely biological, or merely human. Accordingly, the argument goes, when Darwinians claim that human culture, religion included, reflects nothing more than the biologically propelled efforts of human beings to survive as best they can, they undercut that warrant; for science, too, including Darwinian theory, is part of human culture. Thus Haught can claim that, in naturalizing culture, evolutionary psychologists "sabotag[e] the authority of Darwinism itself." This supposedly damaging conclusion about science, however—that its theories reflect nothing more (transcendent) than the biologically (and otherwise) shaped ideas and activities of mere mortal humans—could also be seen as among its most solid justifications. For, in accord with the pragmatist views of scientific knowledge outlined in chapter 1, it could be maintained that what gives the cultural form (or set of ideas and practices) we call science its epistemic authority is not the putatively transcendent truth of its theories but the fact that its models of the operations of the material-physical world enable us to predict, shape, and intervene in those operations more effectively in relation to our purposes. A thoroughly naturalized account of scientific knowledge, then, does not undermine its authority. Rather, it offers a tough-minded, hard-nosed explanation of the sources and workings of that authority.

At this point, a certain irony may be clear to readers familiar with the recent "science wars." For it is just that pragmatist and, to some ears, deflationary account of scientific knowledge and method—that is, as human, cultural, opportunistic, and sustained largely by practical effectiveness—that has emerged in recent years from, precisely, the naturalistic study of science. The irony here is the close resemblance of that project, the pursuit and very idea of which have been deplored and assailed by a number of scientists and philosophers, to the New Naturalistic study of religion, a project that many of those same scientists and philosophers applaud and promote.[28] The similarities between these two naturalizing enterprises, sociological-historical studies of science on the one hand and cognitive-evolutionary studies of religion on the other, are extensive. They include inspiration and method, which in both cases are determinedly empirical, comparative, and impartial; and they also include outcome and reception, which in both cases involve radical challenges to traditional views of a highly valued realm of ideas and practices greeted by widespread misunderstanding, misrepresentation, and outrage. Significantly here, defenders of traditional views in both cases (of science and of religion) have perceived efforts to explain dispassionately how those ideas and practices originate and operate as ignorant, personally motivated attacks on them. And, in both cases, various defenders of traditional views have sought to discredit and dismiss the naturalizing projects in question as logically self-refuting.[29]

The absence of due reflexivity on the part of those offering cognitive-evolutionary explanations of religious belief makes them vulnerable to the theologian's taunt of "*tu quoque*." And, as in Haught's critique examined above, the taunt may be attended by the charge of a supposedly logically entailed self-refutation: You, too; your faith in science is just as merely naturally caused and therefore just as deluded as you say our religion is. A series of moves along such lines is deployed by Paul J. Griffiths in his review of Boyer's *Religion Explained.* Griffiths writes:

> Boyer wants to provide an evolutionary explanation for the plausibility of religious belief and practice, and in so doing to show that it [religion, presumably] should not be taken seriously in its claims about the way things are. But such explanations ought to be applicable to Boyer's own views, since he claims that everything about our cognitive life can be explained by appeal to our evolutionary history. What is it, then, about evolutionary selection that makes Boyer's views (his physicalism, his evolutionism, his touching faith in science and its high priests, his apocalyptic enthusiasm for what science can now do) probative? Boyer does not say. His views are, apparently, exempt from the very process of investigation they require. The whole program is thus performatively incoherent, propounding as it does a method of analysis that ought to be applicable to all claims and arguments, and yet exempting itself from that very process.

This supposed demonstration of Boyer's performative incoherence—or self-refutation—is, however, imperfectly executed.[30]

First, Griffiths encumbers Boyer with the polemical aim of wanting to use evolutionary explanations of religious beliefs to "show that [religion] should not be taken seriously in its claims about the way things are." Boyer does present some familiar—including Christian—beliefs as manifestly counterfactual and remarks that scientific explanations of natural phenomena have always triumphed over religious accounts. But he does not argue that those beliefs or accounts should be rejected *because* their plausibility can be explained in terms of evolved cognitive dispositions. Moreover and more crucially, the observation that some set of beliefs, whether Christian teachings or naturalistic accounts of religion, can be given, in that sense, "an evolutionary explanation" is not equivalent to the claim that those beliefs, teachings, or accounts are worthless or "should not be taken seriously." No doubt, as Griffiths suggests, an explanation in terms of evolved cognitive dispositions could be given for the origin of Boyer's own views, including what Griffiths describes as "his physicalism, his evolutionism, [and] his touching faith in science," and such an "evolutionary explanation" could also be given for the plausibility of those views for Boyer and whoever else accepts them. The existence or possibility of such explanations, however, does not diminish, much less destroy, the intellectual interest or value of those views. This is not to say that Boyer's particular evolutionary accounts of religion (or his unabashed scientism) could not be challenged on some relevant intellectual grounds, for example, conceptual, empirical, or methodological. (I have, of course, sought to challenge them in such ways in this book.) It is to say that Boyer's and comparable evolutionary accounts of religion cannot be dismissed on the spurious ground of performative incoherence.

Insofar as Griffiths's argument incorrectly presumes an equivalence between explainable-by-evolution and not-to-be-taken-seriously, his charge against Boyer misfires. In accusing Boyer's cognitive-evolutionary explanation of implicit self-exemption, however, Griffiths does hit his target. Boyer's insistence, in *Religion Explained*, on the universality and uniformity of evolved cognitive dispositions, might have led him to recognize that the attractiveness or plausibility of evolutionary (and all other naturalistic and scientific) theories is, like that of religious beliefs, subject to explanation in terms of the operation of, among other things, such dispositions. But it evidently did not. On the contrary, Boyer concludes his book with the claim, echoed by a number of other New Naturalists, that, given our evolved cognitive dispositions, religious beliefs are "natural" for humans while scientific ideas are "unnatural."[31] We shall examine that claim and the sharp contrast it involves in the next chapter. What is significant here is that Griffiths and other theistic critics are correct in observing both that naturalistic explanations are themselves subject to naturalistic explanation and also that programmatic naturalists do not commonly acknowledge the fact.

Each of the two sets of points argued above needs emphasis. In their cognitive-evolutionary accounts of the wholly "natural" springs of human, including religious, beliefs, the New Naturalists are not self-refuting. But in their implicit—and sometimes explicit and elaborated—self-exemption, they are self-ignorant.

CHAPTER FIVE

Reflections: Science and Religion, Natural and Unnatural

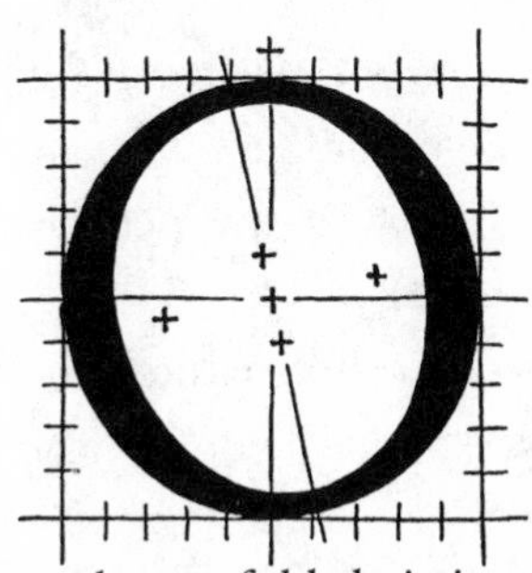

One of my central points in this book is that there are better and worse ways of pursuing the naturalistic study of religion. There are also, I think, better and worse ways to promote the specifically cognitive-evolutionary accounts of religion discussed here and to unfold their intellectual implications. These different ways of reflecting on the New Naturalism and on the relation of science to religion more generally will be my focus in this final chapter. I begin by considering more closely a term and concept that has been central throughout, that is, "naturalism."

MODES OF NATURALISM

The current theoretical wisdom, shared by many philosophers of science and theologians, is that naturalism comes in two forms, "methodological" and "metaphysical."[1] *Method-*

ological naturalism, often equated with the practices of science, is the principled exclusion of appeals to supernatural beings or forces from a certain class of investigations and theories that we call, partly on that account, "scientific." *Metaphysical* naturalism, often associated with atheism, is the view that there is no realm of being beyond, and no entities or forces other than, "the material," "the physical," or, as it is sometimes said, "that which is in nature." But, it may be asked, what do we mean by any of these terms? Or, to put it another way, how can we specify neutrally, without metaphysical assumptions and thus just going around in circles, that which we want to indicate as *the ontologically complete*? This is a lingering question in debates over the propriety or limits of the naturalistic study of religion. Obviously there can be no simple answer to it, but we may approach the question here from a variety of angles.

The distinction between the two forms of naturalism, methodological and metaphysical, figures crucially in a range of current controversies involving science and religion. For example, theological critic Paul J. Griffiths, who, as we have seen, charges scholars who promote naturalistic approaches to religion with "performative incoherence," argues that, since such scholars evidently think naturalism "is true" but have no empirical grounds for thinking so, they are making metaphysical claims just like the theologians whose claims about religion they reject for just that reason.[2] But, it may be observed in reply, naturalism pursued merely as a method is not the sort of thing that can be thought either true or false. Rather, like using low-octane fuel or following a low-fat diet, the minimalism or self-restraint that defines it can only be thought more or less appropriate for the purposes at hand. Accordingly, it could be argued, the promotion of methodological naturalism in religious studies harbors no metaphysical presuppositions, makes no metaphysical claims, and is therefore not open to the *tu-quoque* charge or charge of self-contradiction.

The distinction between metaphysical and methodological naturalism also figures in replies to proponents of Intelligent Design who charge that the decision by evolutionary theorists to exclude extra-natural agency and purpose from their explanations of biological phenomena is arbitrary, based just on personal belief or disbelief. Here defenders of evolutionary theory may and do point out, first, that the exclusion of appeals to extra-natural agency or purpose has been practiced as a matter of method by theistic as well as atheistic scientists and, second, that it reflects not only longstanding considerations of conceptual economy but also the ongoing observation of the practical reliability of the explanations so constructed.

Several points must be added, however, to the philosophical wisdom just outlined. One is that the distinction between the metaphysical and methodological forms of naturalism is not always clear. Thus it may be observed that, at the apparent limits of scientific knowledge, notably in cosmology and contemporary mathematical physics, the exclusion—and implied denial of the possibility of the existence—of entities or forces other than those currently comprehended by natural science amounts both to a metaphysical claim and to a possibly significant intellectual confinement. It appears, in other words, that, with regard to crucial assumptions at such ontological limit points, methodological naturalism *becomes* metaphysical naturalism and, moreover, may close the doors to what would otherwise be recognized as properly scientific inquiry.[3] Of course, the existence of metaphysical assumptions in the theoretical practices of science does not give an epistemic boost to any specific theistic or religious claim. No theological doctrine becomes more credible because some program operating in the orbits of established science involves assumptions that cannot be tested empirically. But the existence of such crucial assumptions in science does put both theoretical positions, the naturalist's rejection of "spooky forces" as much as the theologian's affirmation of them, equally outside empirical falsification. Under those limit conditions, both are, in effect, matters of ontological taste.

Two other forms of slippage between method and metaphysics are important here. One is significant for the relation between science as a doctrine concerning the nature of the universe and such other ontological doctrines as deism, pantheism, or Buddhism. Here a slippage occurs when ontological claims about the structure and contents of "reality" or "the world" are made on the basis of current accounts in the natural sciences.[4] The second type of slippage is significant for the relation between science as an intellectual pursuit, with associated goals and methods, and such other intellectual pursuits as philosophy, literary studies, or the humanities disciplines more generally. Here the slippage occurs when general epistemological claims (for example, what kind of knowledge is "genuine" or how all knowledge should be pursued) are made on the basis of the successes of the various natural sciences in meeting their specific epistemic goals. These two forms of slippage, from a due appreciation of the achievements of methodological naturalism either to manifestly metaphysical claims about what is true of the universe or to peremptory epistemological claims about the "best" or "only" way to arrive at "genuine" knowledge, figure centrally in the triumphalist views and dubious integrationist projects that I described earlier as scientistic. I would stress here that such views and projects are not confined to scientists or even characteristic of them. On the contrary, working scientists appear more likely to find generalized celebratory descriptions of science by philosophers and other nonscientists remote from their own practices and, similarly, to find the idea of an ultimate integration of all knowledge puzzling or gratuitous.

In most current definitions and invocations of naturalism, "nature"—along with what is and is not "in" it—tends to be treated as an unproblematic concept or a self-evidently given realm of being. But, of course, neither is the case. Without rehearsing the long history of the term in Western thought and discourse, we may recall the very length of that history, originating with Greek *phusis* and Latin *natura*, and the multiplicity of meanings that would be yielded by an analysis of even its current formal usages.[5] Notably, while in many scientific and philosophical contexts "nature" is taken as equivalent to everything that exists or what I referred to above as "the ontologically complete," it is also often distinguished from and contrasted to one or more other realms or modes of existence. Thus the "natural" is commonly distinguished from the manmade, the social, or the cultural, and also, as in contrasts between what is (actually) in nature as opposed to what is (merely) in our thoughts, from the intellectual or the mental. In many formal as well as informal discourses, then, "nature" is not always equivalent to everything-that-exists. Complicating matters further is the fact that "nature" and "natural" are commonly used both to praise and to blame. Thus we often distinguish normatively between the "natural," meaning the exemplary or proper, and the "unnatural," meaning the deviant or improper. But, in a usage that will concern us below, distinctions are also sometimes drawn between the "natural," meaning the ordinary, expected, or unremarkable, and the "unnatural," meaning the extraordinary or rare and, for that reason, especially valuable.

Finally, bringing us back to a key perplexity throughout this book, we distinguish the "natural" from what we call the "supernatural," meaning above or beyond "nature," itself meaning . . . well, meaning *what*, given all the above? References to the supernatural in contemporary studies of and disputes over religion typically take for granted a certain more or less modern (Enlightenment, scientific) conception of "nature." Indeed, the two terms, "nature" and "supernatural," appear to a large extent mutually dependent and reciprocally specifying, at least in contemporary Western discourse. That is, while part of what we mean by "supernatural" is other-than-natural, the reverse is also true: our conceptualization of "nature" involves the idea of other-than-supernatural. What makes this tight lexical-conceptual relationship a problem for the study of religion is that many ancient languages and cultures did not have those terms, conceptions, or related contra-distinctions and many non-Western languages and cultures still do not.[6] These multiple linguistic, historical, and cultural divergences and semantic-conceptual instabilities have considerable bearing on how we study, discuss, and generalize about religion. It will not do to brush consideration of them aside as either a stalling tactic of crypto-theological traditionalists or a concern only of pedants and "postmodernists."[7] It will not do because much current punditry on the topic of religion as issued from credentialed scientists and philosophers is rendered vacuous—substantively empty—by just such considerations.

Clearly, whatever else it is, "nature" is a notion—an idea, an abstraction, a human construct. We (Westerners, scientists) keep constructing nature (*natura*, *die Natur*; and so forth) collectively out of our intersubjectively communicated experiences of publicly available phenomena, the latter as distinct from such private, idiosyncratic, and/or inexpressible experiences as dreams or mystical visions. (Given individual perceptual quirks, not to mention inter-communal differences, there are probably limits to the number of persons who can experience *any* phenomenon in what could be described as the same way; but, by conventional definition, what is observed or experienced *only* by a single individual is not included in what we generally understand as "nature.") At the same time, each of us keeps constructing our own personal cosmos out of virtually everything that we experience, here including our private dreams and, as may happen, our individual visions or intuitions of the unity, duality, or divinity of the universe. Nature—the real, the given, the ultimate referent of what's-out-there or what-there-is—is a collective construct. As such, it is shared by the members of an epistemic community that may be, though necessarily spatially specific and temporally finite, nevertheless quite extensive and longstanding. But our personal cosmologies or universe-constructs are highly individuated and, as such, unique to each of us.[8] I do not refer to these subjective cosmologies as either "convictions" or "beliefs" because I wish to avoid the suggestion that they are verbally articulated propositions "about" "the world." Rather, in accord with the views of cognition and belief presented in earlier chapters, what I am concerned with here are heterogeneous and often minimally articulated assemblages—shifting patchworks of ideas, images, schemes, and stories—drawn from the many domains and strata of our experience.

Most of us, I think, including most scientists, would agree that, as Hamlet puts it, there are more things in heaven and earth than dreamt of in our philosophies—or, as we may gloss the much-cited passage for our present purposes, there are more types of entities and forces in our individual experiences of the universe than set forth in current scientific models of the shared phenomenal world. Some people, however, aspire to be full-time metaphysical naturalists and do not permit anything to enter their personal cosmologies, or acknowledge that there *is* anything in them, except what they have put there on good scientific authority.[9] These include a number of evangelical atheists: that is, full-time metaphysical naturalists convinced that the way they conduct their own cognitive-imaginative lives is the way all other people should conduct theirs. I return to such convictions below. First, however, it will be useful to consider some sharp contrasts drawn between science and religion on the basis of New Naturalist explanations of religion and related understandings of human cognition.

NATURAL RELIGION / UNNATURAL SCIENCE

In the final chapter of *Religion Explained*, Pascal Boyer begins a discussion of the relation between religion and science by rejecting a simple opposition between the two in favor of quasi-nominalist characterizations of each. "It is by no means clear," he writes, "that there is such a thing as 'religion' in the abstract." Rather, he continues, "[t]here are many mental representations entertained by people, many acts of communication that make them more or less plausible, many inferences produced in many contexts." He goes on to make parallel observations of science: "Science too is a cultural thing, that is, a domain of mental representations that happen to be entertained by a number of human minds. There is no science as such but rather a large set of people with particular activities, a particular database that is stored in a particular literature, and a particular way of adding to or modifying that database." The apparent nominalism and symmetry soon dissolve, however, in favor of a familiar story of epistemic triumph and routing. In the West, Boyer writes, a "monopolistic doctrinal religion" (evidently Roman Catholicism) "made the crucial mistake of meddling in empirical statements of fact. . . . In every instance where the Church has tried to offer its own description of what happens in the world *and* there was some scientific alternative on the very same topic, the latter has proved better. Every battle has been lost and conclusively so."[10] We also return to a familiar generalized and essentialized concept of science and, with it, to the assertion of its fundamental opposition to religion, similarly generalized and essentialized: "Science showed not only that some stories about the formation of planets were decidedly below par but also that there was something dramatically flawed *in principle* about religion as a way of knowing things and that there was a better way of gathering reliable information about the world." Boyer concludes his discussion by reframing the opposition as a contrast between the inherent attractiveness of religious concepts versus scientific thinking for the human mind. Given our evolved mental dispositions, he writes, religious concepts are "a *likely* thing" for us whereas scientific activity, being cognitively "unnatural," is an "*unlikely*" thing theoretically and in fact quite rare among humans. To support the terms of this contrast, Boyer cites a recent essay by Robert McCauley titled "The Naturalness of Religion and the Unnaturalness of Science," which itself echoes the arguments (and title) of a book by British biologist Lewis Wolpert, *The Unnatural Nature of Science*.[11] These shared views and duplicated arguments are worth our attention here.

McCauley remarks at the beginning of his essay that it is provoked by scholars who maintain that, because religion is not—or not simply—a natural phenomenon, its study requires methods other than those of the natural sciences. Seeking to turn the tables on such arguments, his intention is to demonstrate that religion is, on the contrary, something supremely natural while it is actually science that is unnatural. As is often the case with polemical table-turnings, however, the reversal here does not come off altogether smoothly. McCauley's argument, as he lays it out, consists of a series of strongly contrastive characterizations appealing to apparently straightforward observations supplemented by references to historical and experimental evidence. Thus from the fact that religion is found in all times and cultures, he argues that we may conclude that it requires nothing but the universals of human nature to spring up; conversely, given the historical and cultural rarity of science, we may conclude that it is essentially contrary to human nature. Or, later, he observes that inasmuch as science requires literacy, complex social arrangements, educated elites, and technical means for preserving and transmitting knowledge, it is fundamentally "cultural" while, conversely, inasmuch as religion requires nothing but "basic cognitive abilities," it is natural. Or again, the fact that religious concepts are easy to learn and to remember and quickly acquired even by young children indicates that such concepts conform to innate intuitions, while the fact that scientific concepts are hard to learn and take specialists years to master is evidence that they are counterintuitive and demand exceptional forms of cognitive discipline. These contrasts are in some ways plausible-sounding; they draw on familiar observations; and they are presented by McCauley with a string of references to the psychological literature. The distinctions and alignments on which they are based, however, involve crucial conceptual oversimplification and historical obliteration.[12]

For one thing, it is not clear that comparable matters are being compared here. Thus, at the simplest level, we may ask what exactly it is in religion or religious concepts that children acquire so easily and in science or scientific concepts that most people never come to master. To be sure, many children who can recite their prayers with ease would have considerable difficulty explaining Einstein's Unified Field Theory. But many children who can recite the multiplication table or give the chemical formula for water at the drop of a hat would have considerable difficulty explaining the Doctrine of the Trinity. What appears to be the case, but is distorted here in the service of a strained contrast, is that certain concepts and verbal routines, religious, scientific, philosophical, and others, are acquired readily while other more complex or sophisticated concepts and formulations, again from any and all domains of thought, require a specialized education and long apprenticeship for their mastery. Second and more fundamentally, in McCauley's essay as in Boyer's and Wolpert's books and other current invocations of the natural-unnatural opposition, the contrast between science and religion requires the representation of each as a monolith and the definition of each in artificially broad, artificially narrow, or otherwise strained ways. For example, while McCauley evidently includes in "religion" everything from Neanderthal burial practices (one of his examples) to the catechisms taught in local parishes, he insists on a narrow, historically and culturally quite specific, understanding of "science," which of course begs the question of science's alleged historical and cultural rarity.

Here as elsewhere in current writings on these issues, sharp distinctions and strong contrasts between science and religion require our forgetting quite a bit of recorded human history, notably the extensive historical overlaps and continuities between the two. These include, for the better part of the past millennium, close intellectual as well as institutional ties between Western science and the Catholic Church. Historians of the subject remind us that much of what we now call science—pursued as "natural philosophy"—was developed in medieval universities originally based in monastic orders and that scientific pursuits remained theologically oriented long afterward. As late as the eighteenth century, nature was studied systematically by, among others, Isaac Newton on the assumption that it embodied divine purpose and with the aim of revealing just how it did so.[13] A number of familiar ideals in science, such as the unity and perfectibility of knowledge, appear to be the fairly direct heritage of Christian doctrine, transmitted through the medieval universities and extended by Enlightenment and, later, "evolutionary" narratives of human rationality, development, and, again, perfectibility. These and other norms, ideals, and assumptions shared by Western science and Western monotheistic religions seem to reflect more general human tendencies, cognitive and other: for example, the tendency of people everywhere to construct teleological, meliorist narratives or to suppose that a strong male presence is required for important works of the mind or spirit.[14]

The generality of cognitive tendencies among humans and the continuity of cognitive processes in the practices of religion, science, and everyday life are both acknowledged and denied by New Naturalists. Thus, while Boyer emphasizes throughout *Religion Explained* that religious persons are not essentially different from nonreligious ones in essential cognitive functions,[15] he also maintains, as in the passages quoted above, the exceptional cognitive and motivational character of scientists. The crucial point, he goes on to argue, is that, because of their special training, disciplined individual efforts, and the unique normative system that defines their community, scientists come to act in ways that supersede their species-characteristic cognitive dispositions and impulses. That may be true. But the same could be said of, among others, Buddhist monks, classical scholars, and Oxford-educated analytic philosophers, each of whom, given their special training, disciplined individual efforts, and the distinctive normative systems that define their respective communities, could (and often do) make the same claims about their transcendence of ordinary human limits, cognitive and other. Similarly, while McCauley acknowledges that scientists "exhibit the same cognitive biases and limitations that other human beings do," he goes on to maintain that scientists "get around" such biases and limitations because they have special "tools (such as literacy and mathematical description)" and because institutionally established norms encourage them to "seiz[e] opportunities to criticize and correct each other's work."[16] The tools and norms that McCauley invokes here are certainly significant in limiting the effects of scientists' cognitive liabilities. Their operation, however, is not as simple as he implies, nor their effectiveness as decisive.

Among other epistemically significant cognitive tendencies that scientists share with nonscientists are animism, anthropomorphism, overgeneralization, essentialism, reification, hypertrophy, binary thinking, hierarchical thinking, linear-causal thinking, teleological thinking, and a tendency to divide the social or intellectual world into communities of "good/right-us" and "bad/wrong-them." A number of these shared cognitive tendencies are illustrated in the conduct of the two contemporary intellectual projects, one scientific, the other theological, examined in previous chapters here, that is, the New Naturalism and the New Natural Theology. As suggested in chapter 1, scientists, both individually and as members of epistemic collectives, may rationalize failed predictions in regard to current scientific theories in ways that resemble the belief-preserving cognitive activities of millenarians in regard to failed prophecies. More generally, it has been observed that, just as epistemically beneficial communal norms (accurate observation, precise statement, mutual criticism, and so forth) are established and sustained through what McCauley calls "the institution of science," so also are more ambivalently operating theoretical assumptions, shared conceptual and discursive idioms, and related communal habits of perception and classification. The cognitively conservative—belief-preserving, innovation-discouraging, paradigm-hardening—operations of these latter shared practices among scientists have been documented and theorized by historians and sociologists of science for close to a century.[17] Accordingly, most contemporary philosophers of science reject the idea or claim that science is automatically, dependably, or even somewhat miraculously "self-correcting."

I noted above that McCauley insists on a tendentiously narrow definition of science. To appreciate its "rarity," he writes (quoting Wolpert), we must not "confuse" science with "technology." He explains: "The crucial point is that the practical orientation of technology and the abstract theoretical interest in understanding nature that characterizes science are not the same aims." Or again, he writes: "Science is finally concerned with understanding nature for its own sake and not merely for its effects on us."[18] But the admonition not to confuse science and technology, though familiar, is not so easy to heed. On the contrary, distinguishing them at all requires some significant retrospective tinkering. Most of the specialized pursuits we now associate with Western science, including anatomy, botany, chemistry, and physics, developed in close conjunction with technical problem-solving in such perennial human activities as healing, agriculture, navigation, and warfare. A tradition and image of gentlemen investigators interested in understanding the workings of nature "for its own sake" emerged in the seventeenth century, largely in the science academies of England and Europe. But the conjunction of such investigative pursuits with practical activities continued and, with the dominance, since World War II, of large-scale scientific ventures funded mainly by governmental, industrial, and commercial agencies, any effort to mark off a realm of pure science pursued independent of "a practical orientation toward technology" can only be arbitrary and artificial.[19]

The depictions of science required to sustain the unnatural/natural contrast with religion are not only conceptually and historically strained but pose a technical puzzle or "evolutionary riddle" for McCauley and the other New Naturalists maintaining it. For, given their identification of science with a cognitively unnatural "abstract theoretical interest in understanding nature" "for its own sake," the question arises as to how, from an evolutionary perspective, such an enterprise—that is, one with no material advantages or connection to individual interests—could have arisen in the first place among humans and why science has survived at all. Indeed McCauley seems to be led by just such considerations to represent science (as he defines it) as something quite fragile, particularly in competition with religion (as he defines it). He writes: "In the global marketplace of ideas . . . some views have natural disadvantages. Science, with its esoteric interests, its counterintuitive claims, and its specialized forms of thinking, certainly seems to qualify. [Some scholars] . . . hold that science was once lost and had to be reinvented. One consequence of my view is that nothing about human nature would ever prevent its loss again."[20] This sounds rather ominous. There is, however, good reason to doubt that the survival of science, non-tendentiously defined, is really so precarious. Certainly the apocalyptic visions of an imminent return to the Dark Ages are overblown, forgetting the immense practical benefits, individual and communal, attached to scientific ideas, models, and explanations and, in many cases, exaggerating both the global significance of religious fundamentalism and the extent to which participation in traditional religions amounts to simple benightedness.[21] Moreover, given the evolutionary dynamics of what is invoked here by McCauley and represented more generally by other New Naturalists as "human nature," the cognitive springs of science, even at its most "esoteric," "counterintuitive," and "abstract," do not appear all that unnatural or, indeed, all that remote from the springs of religion, non-tendentiously defined.

It seems clear that the array of distinctively human practices and techniques we now call science arose in the course of efforts by our ancestors to solve practical problems of survival and that such practices and techniques were shaped cognitively, as well as culturally and materially, largely by their effectiveness in serving those ends. Rather than technology being, as McCauley and Wolpert represent it, an incidental byproduct of pure science, the reverse seems closer to the truth. That is, what they frame as the defining characteristic of science, "the abstract theoretical interest in understanding nature" "for its own sake," appears to be an offshoot of technology, a byproduct of cognitive capacities and tendencies that were naturally selected because they served more practically oriented activities. McCauley and Wolpert's pure science is thus what Boyer calls, in regard to religious concepts and practices, "parasitic": a type of activity that emerges and persists among humans not because it confers fitness benefits itself but because (in Boyer's terms) it "recruits" or "piggybacks on" cognitive and other faculties or impulses that conferred such benefits in the course of our evolution. Like other such activities (performing music, playing chess, having sex for its own sake, and so forth), the pursuit of science for its own sake exercises cognitive faculties and responds to bodily impulses that may or may not continue to serve particular fitness-related functions but the exercise and satisfaction of which are pleasurable in themselves. In the case of pure science, such pleasure may be derived from the construction of explanatory models of features of the world just for the sake of producing and contemplating them, quite apart from any practical applications they might yield.

Interestingly, a view of science as "parasitic" in just this way has been advanced by developmental/cognitive-evolutionary psychologist Alison Gopnik. "Science is successful," she writes, citing recent experimental findings, "because it capitalizes on a more basic human cognitive capacity," what she calls "the theory formation system drive." According to Gopnik, the fulfillment of that drive yields the deep satisfaction that humans, including young children, characteristically experience in the production of good explanations. She remarks: "Science is thus a kind of epiphenomenon of cognitive development. It is not that children are little scientists [a view that Gopnik advances elsewhere] but that scientists are big children," getting, in effect, a rush or high from the fulfillment of an elementary drive (she compares it explicitly to sexual pleasure).[22] Much in Gopnik's account can be disputed. It is rather heavy on dubiously postulated drives and mental systems, and the "orgasmic" fulfillment-high that she claims is produced specifically by the production of good proto-scientific explanations is not clearly distinguishable from the kinds of satisfaction elicited by the successful completion of any strenuous creative or intellectual venture or, indeed, by the successful execution of any difficult physical, for example, athletic, feat. But Gopnik's observations are suggestive here and make clear that what Boyer, McCauley, Wolpert, and others allege to be the cognitive unnaturalness of science is by no means self-evident to all cognitive scientists. Moreover, and significantly for the discussion below, her observations illustrate the possibility of a thoroughly naturalistic cognitive-evolutionary account of the most highly valued aspects of science—and highly valued traits of its practitioners—and give some idea of how unsettling such an account might appear.

The differences between science and religion, each duly historically defined and comprehensively indicated, remain important; and, of course, the stakes, both political and intellectual, in distinguishing them appropriately are sometimes very high. But here as elsewhere, the better way to go is careful delineation and discrimination rather than tendentious characterization or exaggerated contrast—better in both the political short run and the intellectual long run. Scientists, and academics and intellectuals generally, have reason to be wary, now as ever, of the political exploitation of public ignorance. Due public regard for scientific knowledge, however, is not secured by sheer self-celebration or by claims of a radical difference in cognitive kind and epistemic value between science and every other intellectual pursuit.

THE NEW NATURALISM AND THE TWO CULTURES

The operation of a strong Two Cultures ideology, with its familiar intellectual provincialisms, inter-disciplinary hostilities, and mutual caricatures of "scientists" and "humanists," has, over the past fifty years, come to dominate the Anglo-American academy. I have detailed its manifestations and effects elsewhere.[23] What is significant here is the perpetuation of that ideology by many practitioners and promoters of the New Naturalism. The caricatures here include a constellation of routinely disparaged and often conflated antagonists or supposed antagonists: theologians, scholars of religion, humanities scholars, cultural anthropologists and practitioners of other interpretive social sciences, and, to be sure, "postmodernists." Each of these groups is represented as hostile to "cognitive" or "evolutionary" explanations of religion out of one or another intellectual weakness and/or moral flaw: a vested theological or theistic anti-naturalism, a sentimental view of human uniqueness, a personally or politically motivated blank-slate conception of human nature, or a cynical view of science and truth.

These conflations and dubious characterizations may be found at various points in Daniel Dennett's *Breaking the Spell*, which celebrates current cognitive-evolutionary accounts of religion against what Dennett represents as a history of academic resistance to, and indeed a "taboo" against, naturalistic explanations of religion and all other human phenomena. In an arresting passage, he writes:

> The ardent anti-Darwinians in the humanities and social sciences have traditionally feared that an evolutionary approach would drown their cherished way of thinking—with its heroic authors and artists and inventors and other defenders and lovers of ideas. And so they have tended to declare, with desperate conviction but no evidence or argument, that human culture and human society can only be interpreted and never causally explained, using methods and presuppositions that *are completely incommensurable with*, or *untranslatable into*, the methods and presuppositions of the natural sciences.[24]

The intellectual mischief here is substantial. Since Dennett supplies no names or direct quotations to give substance to this purported report, it is hard to know who or what he is talking about or, of course, to challenge the broad characterizations with specific counter-evidence.[25] To virtually anyone associated with any of the fields in question, however, what Dennett represents as the "desperate conviction" and traditional fears of "ardent anti-Darwinians in the humanities and social sciences" will appear as a distortion of the views and motives of those who have maintained the intellectual interest and value of interpretive and related non-natural-science approaches in the various non-natural-science disciplines.

As I noted in chapter 2 above, the exclusivist position that Dennett indicts here—"*only* interpretation, *never* explanation"—has been maintained by scholars of religion who invoke what they see as the unique and, in effect, hallowed character of religion to argue against the very possibility of naturalistic approaches to their subject. But this severely confining view of the study of religion and a comparably rigid resistance to invocations of biology in connection with human culture among some humanities and social-science scholars are significantly different from the more broadly representative views with which Dennett here misleadingly conflates them.

Among those who study religion, including cultural anthropologists and cultural historians, the more typical view is that an adequate understanding of the religions of the world requires the sorts of historical, comparative, and interpretive approaches associated with the humanities *as distinct from*, but not *to the exclusion of*, the experimental, quantitative, explanatory approaches associated with the natural sciences. It is not the case, as Dennett asserts, that such a position has been maintained "with no evidence or argument." There are extensive arguments—epistemological, pedagogic, and other—associated with such views.[26] If Dennett denies their existence here, it may be because he is convinced they cannot be proper arguments and discounts them in advance. As for evidence, in view of the ongoing pursuit of historical, comparative, and interpretive approaches by scholars in the fields just mentioned and the ongoing invocation and appropriation of their work by other members of the intellectual community, it is not clear what further evidence is needed—or what other type could be demanded or supplied—to establish the interest and value of those approaches.

Comparable positions—that is, views to the effect that interpretive and other non-natural-science approaches are crucial in achieving the intellectual aims of their particular fields of study—can be found among art historians, literary scholars, musicologists, philosophers, and various social scientists as well as among those who study religion. Many anthropologists, for example, maintain the distinctive value of accounts of particular cultures from the viewpoint of participants as distinct from those produced from the outsider viewpoint of observers—though, again, not typically to the exclusion of the latter accounts.[27] More generally, scholars in most humanities fields are persuaded, for good reasons, that there are important intellectual purposes distinctively secured by interpretive, comparative, historical, and related humanities-identified approaches to such specifically human activities as art, music, literature, and philosophy—approaches derided here by Dennett as "their cherished way of thinking—with its heroic authors and artists and inventors and other defenders and lovers of ideas." These scholars might very well claim that the value of such approaches and related practices is, as Dennett writes in italics, "*incommensurable with*" that of natural-science explanations: not, however, as Dennett seems to think, in being immeasurably superior to scientific explanations but, rather, in the sense that the products of such approaches and practices—for example, detailed individual descriptions, historical placements and contextualization, cross-generic or cross-regional comparisons, or textual and iconographical exegeses—are not usefully assessed by the same measures as natural-science explanations. The existence of intellectual aims and purposes other than those associated with the natural sciences and, accordingly, of other marks and measures of intellectual value is a possibility that Dennett has evidently not entertained. Many members of the intellectual and academic community, however, including many scientists and other philosophers, would agree that the characteristic methods and products of the humanities disciplines are, as Dennett puts it with evident amazement or indignation, "*untranslatable into* the methods and presuppositions of the natural sciences." But they would find it hard to see why he finds the idea so stupefying aside from the fact that it is at odds with his own parochial view of knowledge.

All the methodological positions described above involve an affirmation of the distinctive value of interpretive approaches to the study of human phenomena in fields that have, as their intellectual aims, forms of understanding, insight, connection-making, and reflection that are not characteristically sought by the natural sciences as such. None of these methodological positions, however, issues from what Dennett describes as a fear of evolutionary approaches, and the totality of them does not amount to what he claims has been a "taboo" on the pursuit of naturalistic explanations of human phenomena. Musicology has maintained no taboo against psychoacoustic and information-theoretical explanations of the effects and interest of music. Art history has maintained no taboo against psychological, sociological, and economic explanations of the emergence and reception of artistic styles.[28] These and comparable naturalistic approaches in classical studies, literary studies, and other humanities disciplines have long been methodological staples in the fields in question, though usually combined with more distinctively humanities-based approaches. Most significantly here, the acknowledged dominance of humanistic approaches in the academic field of religious studies per se has not prevented the treatment of religion as a natural phenomenon in many *other* academic fields; nor has it prevented the production of important naturalistic accounts of religious beliefs, practices, and institutions by ethnographers, archeologists, philologists, historians, psychologists, sociologists, and social theorists for the past two and half centuries.[29] Most of these accounts of religion, certainly the ones written since the late nineteenth century, are informed to one degree or another by Darwinian concepts, and none of the latter involves a rejection either of Darwinian or neo-Darwinian theory as such or of its relevance to the understanding of human behavior and culture.[30]

So what "taboo," exactly, is Dennett talking about? And who, exactly, are the "ardent anti-Darwinians in the humanities and social sciences" whose views so exercise him? The answer to the first question is that, although there has been significant resistance to the naturalistic study of human phenomena in some quarters of the academy, there is not now and never has been a wholesale ban on or fear of such study. Dennett seems to have simply whipped up a bit of sensationalist intellectual history here.[31] The answer to the second question is more elusive, but the academics toward whose views Dennett gestures here are probably not anti-*Darwini-*

ans (a term that evokes hide-bound creationists holding forth in university classrooms) but scholars in the humanities and social sciences who challenge the claims or note the limits of various evolution-invoking approaches in their particular fields of study. They may be philosophers who reject the suggestion that moral theory can be replaced by sociobiology. They may be literary scholars unimpressed by the claim of self-described literary Darwinists to determine the meaning and explain the appeal of celebrated novels on the basis of the mammalian mating-strategies that can be identified in their plots.[32] To be sure, there are Two Cultures ideologues in all academic venues and one may encounter facile rejections of an ill-understood "Darwinism" by scientifically ignorant scholars and students in the humanities and various social-science fields.[33] But literal "anti-Darwinians" of that kind should be distinguished from relevantly informed scholars and social scientists who challenge, on recognized intellectual grounds, various assumptions, aims, or claims of sociobiology or evolutionary psychology. They should also be distinguished from scholars who resist, for equally intellectually respectable reasons, the uncritical importation into their own disciplines of undigested concepts, terms, and methods from nominally Darwinian approaches in other disciplines. For what is characteristically resisted in these latter instances is not the possibility of naturalistic, biological, or evolutionary explanations as such, but the claim of natural-scientific authority for crude, glib, pedestrian, inert, or otherwise dubious intellectual ware. It may be added that the indiscriminate endorsement of such evolution-invoking projects and productions in various humanities fields by consilience-minded scientists contributes nothing to the advancement of scientific knowledge or to its ostensibly sought integration with humanistic learning.[34]

CONCLUSIONS, SYMMETRICAL AND ASYMMETRICAL

Programmatic naturalists such as Dennett and E. O. Wilson maintain that impartial, detached, strictly naturalistic accounts can and should be given of all aspects of human behavior and culture. So described, it is an intellectual program for which one may have much sympathy. One may also note, however, that its current application is itself distinctly skewed. Dennett represents the rejection of naturalistic approaches in the study of religion as a self-deluding spell designed to protect religious beliefs from demystification; and, correspondingly, he represents recent cognitive-evolutionary explanations of religion as the breaking of that spell by a few brave iconoclasts and the triumph, through their efforts, of reason and science over fear and superstition. At the same time, however, like E. O. Wilson, Edward Slingerland, and other promoters of naturalistic approaches in the humanities and social sciences, Dennett brushes aside, as an aberration of "postmodernism," the accounts of scientific knowledge currently issuing from such fields as the history and sociology of science—which is to say, from the violation of a longstanding, quite explicit, and quite effective taboo against the thoroughly naturalistic study of *science*.[35] I noted the irony of this conjunction at the end of chapter 4. In institutional terms, it amounts to a science-exceptionalism that makes everything subject to scientific explanation except science. In individual terms, it amounts to an intellectually vulnerable self-blindness among self-declared naturalists.

Nietzsche suggested that natural science, in its ascetic self-discipline and "faith in a *metaphysical* value, the absolute value of *truth*," was the latest and perhaps last expression of what he called priestly, religious ideals. He also wrote, in a passage especially apt here, of an "*unnatural*" science, namely, "the self-critique of knowledge."[36] Self-criticism has long been seen as central to the practices of science and, in the works by McCauley, Wolpert, and Dennett discussed above, as distinguishing scientific knowledge from the dogmatism of religious doctrine. But the type of self-critique that Nietzsche evokes here, the practice of which would identify a science that was "unnatural" in the sense both of extraordinary and self-transcending, would be that which questioned not merely individual scientific hypotheses but the most fundamental assumptions about the nature, genesis, and value of scientific knowledge. It was just such questioning as pursued throughout the twentieth century by a number of intellectually radical scientists along with historians, philosophers, and sociologists of science that issued in the academic enterprise that we now call science studies. That self-questioning also issued in a widespread rejection of the triumphalist view of science that sustains conceptually simplistic and historically dubious representations of the relations between science and religion.

We have observed that scientists studying religion are subject to the same general cognitive dispositions and liabilities that they identify as natural to the human species and as responsible for some of the central features of religion. As I have emphasized from the beginning, this is not equivalent to saying, nor does it follow logically, that science is "just another belief," like shamanism or Presbyterianism.[37] The system of methodological commitments that defines Western science, notably naturalism, empiricism, and experimentalism, constitutes an extremely efficient apparatus for generating models of the operations of the phenomenal-physical universe that permit us to predict, control, and intervene in those operations with maximal effectiveness and reliability. To the extent that shamanism or Presbyterianism seeks such ends, Western science is better at achieving them. Also, many characteristic norms and practices of Western science, such as the detailed description of investigative procedures, the prompt and open publication of findings, and the honoring of fertile theoretical innovation, are especially effective ways to limit the negative consequences of cognitive conservatism and other liabilities of our humanly shared cognitive dispositions. The efficacy of scientific methods, norms, and practices in these several respects, however, does not make the knowledge thereby produced either the only kind of knowledge there is or the only kind worth seeking. Nor does it make the natural sciences the only locus of desirable cognitive self-discipline among humans.

Scientists share cognitive tendencies, achievements, and limits with nonscientists; religious believers share them with nonbelievers. Although each may put the world together and conduct his or her life in ways that are at odds with or opaque to the other, the cosmology and way of life of each deserves minimally respectful acknowledgement from the other. Such acknowledgment would not mean accepting ideas one finds fantastic or claims one knows are false. And of course it would not mean approving practices that one knows are confining, maiming, or murderous to oneself or to others. What it would mean is recognizing, as parallel to one's own, the processes by which those cosmologies and ways of life came to be formed.

Not me, says the self-vaunting evangelical atheist. *Tu quoque*—you, too—says the defensive, resentful theist. *Et ego*—I, too—says the reflexive, reflective naturalist.

Notes

CHAPTER 1. INTRODUCTION

1. See Festinger, Rieken, and Schachter, *When Prophecy Fails*, 30–31, from which the narrative given here is adapted.

2. Ibid., 31, v. The interest in such movements was widespread at the time. Norman Cohn's groundbreaking study, *The Pursuit of the Millennium*, appeared the following year, 1957.

3. For discussion of such questions, see J. T. Richardson, "Experiencing Research on New Religions and 'Cults,'" and Bainbridge, *The Sociology of Religious Movements*, 137–38.

4. Festinger, Rieken, and Schachter, *When Prophecy Fails*, 37, 169.

5. Later investigators of the episode suggest that Keech and other remaining members of the group were prodded into evangelical statements largely by newspaper reporters and other curious onlookers. See Melton, "Spiritualization and Reaffirmation," 146–57.

6. Some of these salvaging maneuvers are noted in Bainbridge, *Sociology of Religious Movements*, 135–36.

7. The book elicited much commentary in the psychological community and reactions among the general public, including a satirical novel by Alison Lurie titled, rather wonderfully, *Imaginary Friends*.

8. My epistemological scruples here and elsewhere in this book ("*apparently* contrary evidence," "*what strikes other people as* clear disconfirmation," etc.) are not observed by Festinger, Rieken, and Schachter. On the contrary, stalwart empiricists to the end, they describe the phenomenon of belief-persistence by evoking the reactions of someone "presented with evidence, unequivocal and undeniable evidence, that his belief is wrong" (*When Prophecy Fails*, 3). There could be little argument, of course, about whether or not an expected flood had occurred. As indicated by the history of both theology and science, however, an argument can always be had about whether a particular non-event amounts to "unequivocal and undeniable evidence" that some prior conviction is mistaken.

9. See Festinger, *A Theory of Cognitive Dissonance.* For the subsequent fortunes of the theory, see Harmon-Jones and Mills, eds., *Cognitive Dissonance.* The "progress" indicated in the subtitle of this collection ("*Progress on a Pivotal Theory in Social Psychology*") can be seen to have had its static and regressive phases. Statements of the original theory became increasingly dogmatic and work on it by later generations of social psychologists, especially in the 1960s and 1970s, became increasingly repetitive, narrow, and formalistic. (An important exception here is Zygmunt, "When Prophecies Fail," which appeals to historical studies of millenarian movements as well as to experimental research to develop a fine-grained but comprehensive general account.) More recently, Festinger's analyses have been subjected to a series of more or less dubious modifications while the phenomenon of belief-persistence has been given various alternative explanations in line with shifting trends in social psychology: "attribution theory," "self-affirmation theory," "self-accountability," "aversive consequence avoidance," and so forth (see the Harmon-Jones and Mills collection cited above). For a new twist on Festinger's theory (and UFO cults) by way of rational-choice theories of religion, see Bader, "When Prophecy Passes Unnoticed." For a recent effort by a psychiatrist to synthesize cognitive dissonance theory with psychoanalytic theory and neuroscience, see Wexler, *Brain and Culture.* For discussion of some ethical, epistemological,

and other theoretical implications of cognitive dissonance, see B. H. Smith, "The Complex Agony of Injustice"; *Belief and Resistance*, xiv–xv, 148–49; *Scandalous Knowledge*, 154–55.

10. See Lightman and Gingerich, "When Do Anomalies Begin?" for an analysis of what they call the "retrorecognition" of anomalies, that is, the tendency of practicing scientists to discount anomalous findings until after those findings have been given stable explanations in a new conceptual framework.

11. The costs were evidently primarily emotional but, for members who traveled far, left jobs, gave away belongings, and so forth, they were material as well.

12. See Mynatt, Doherty, and Tweney, "Confirmation Bias in a Simulated Research Environment"; Nisbett and Ross, *Human Inference;* Kahneman, Slovic, and Tversky, eds., *Judgment under Uncertainty;* Johnson-Laird and Shafir, "The Interaction between Reasoning and Decision Making."

13. See Gigerenzer, *Adaptive Thinking.*

14. See Harmon-Jones, "Toward an Understanding of the Motivation Underlying Dissonance Effects." Harmon-Jones stresses the adaptiveness, but not the ambivalence, of the persistence of belief.

15. Such matched lists are a recurrent feature of hostile exchanges between atheists and theists.

16. See B. H. Smith, *Belief and Resistance*, 23–51, for elaborations of a number of these points. For important dynamic, ecological, and/or embodied accounts of cognition, see Oyama, *The Ontogeny of Information;* Varela, Thompson, and Rosch, *The Embodied Mind;* Thelen and L. B. Smith, *A Dynamic Systems Approach to the Development of Cognition and Action;* van Gelder and Port, "It's About Time"; Hendriks-Jansen, *Catching Ourselves in the Act;* Hutchins, *Cognition in the Wild;* Núñez and Freeman, eds., *Reclaiming Cognition.* For more recently elaborated views of thought, perception, and/or cognition that stress embodiment and ongoing dynamic interaction, see Melser, *The Act of Thinking;* Noë, *Action in Perception;* Rockwell, *Neither Brain nor Ghost;* Wheeler, *Reconstructing the Cognitive World;* Niedenthal, "Embodying Emotion."

17. For influential examples of constructivist and/or pragmatist views of knowledge and science, see Fleck, *Genesis and Development of a Scientific Fact;* Foucault, *The Archaeology of Knowledge;* Feyerabend, *Against Method;* Goodman, *Ways of Worldmaking;* Rorty, *Philosophy and the Mirror of Nature;* Bloor, *Knowledge and Social Imagery;* Knorr-Cetina, *The Manufacture of Knowledge;* Latour, *The Pasteurization of France* and *Science in Action;* Pickering, *The Mangle of Practice.* For a useful historical survey, see Golinksi, *Making Natural Knowledge.*

18. Fleck, *Genesis and Development,* 27, translation modified.

19. For further discussion of these points, see B. H. Smith, *Scandalous Knowledge,* 1–17, 46–84.

20. While sheer social affirmation is probably insufficient to sustain an otherwise troubled conviction, the significance of *language* in the transmission and preservation of beliefs can hardly be overestimated. Social-linguistic interactions are crucial in forming beliefs, and verbal articulation itself, even without ritual repetition or textual codification, is a powerful force in stabilizing and conserving them.

21. The point is noted in Zygmunt, "When Prophecies Fail." Zygmunt stresses the significance of interactions among members of a social collective (rather than just individual psychological mechanisms in response to other people) both in coping with disappointed expectations and in establishing beliefs to begin with. For a good account of relevant developments in the field of social psychology, focusing on the gradual impoverishment of models of social cognition since the early twentieth century, see Greenwood, *The Disappearance of the Social in American Social Psychology.* For important recent work on social cognition, see Caporael, "The Evolution of Truly Social Cognition."

22. Kafka, "Leopards in the Temple," 93.

23. Brooke, "Natural Theology," 61; see also Brooke and Cantor, *Reconstructing Nature,* esp. chapter 6.

24. See chapter 3 below for references, examples, and further discussion. I discuss this incoherence elsewhere as "scrappiness" (see

B. H. Smith, *Contingencies of Value*, 140–44). The term is meant to suggest that our beliefs are largely in bits and pieces, like scraps, and often at least implicitly conflicting, as in a fight or a scrap.

25. An exception is Pascal Boyer, who rejects the "theologistic" idea that religious beliefs are logically connected (see Boyer, *The Naturalness of Religious Ideas*, 228–30). What Boyer stresses here, however, is not the general heterogeneity of people's beliefs but the view, crucial to his account of religion, that the co-occurrence of various religious ideas is an effect of particular cognitive processes. See chapter 2 below for discussion of Boyer's later work *Religion Explained.*

26. Wallace, *Religion*, 265, quoted in Bainbridge, *Sociology of Religious Movements*, 404. Comparable and especially influential passages can be found in the work of sociologist Peter Berger from the same era; see, e.g., *The Sacred Canopy*, 105–71.

27. Beckford, *Social Theory and Religion*, 70.

28. Fleck, *Genesis and Development*, 28.

29. Beckford, *Social Theory and Religion*, 70.

30. Hinde, *Why Gods Persist*, is an early example of such efforts.

31. R. P. Carroll, *When Prophecy Failed*, 124. The passage is in italics in the original text.

32. For a recent example, see Paul J. Griffiths's review, in the Catholic-oriented journal *First Things*, of a naturalism-promoting book by scholar of religion Russell T. McCutcheon. (The review is specifically titled "*Tu Quoque.*") Griffiths concludes by noting "the extent to which it [McCutcheon's book] exhibits precisely the blindness against which it inveighs [in scholars of religion such as Mircea Eliade]," namely, "being insufficiently aware of the complex connections between [one's] theoretical views and the institutional, political, and economic setting that makes the development of these views possible, and which is in turn supported and validated by these views" ("*Tu Quoque*," 48). Griffiths comments acidly: "But to what forms of hegemony is McCutcheon's [naturalistic] program related? For what institutional unpleasantnesses and oppressions is it both

lackey and paid theorist? McCutcheon doesn't even ask: he seems to think that his work and the kind of work he recommends are immune from such connections." I discuss a similar example of the *tu-quoque* charge by Griffiths in chapter 4 below.

33. For relevant accounts of the grounds (or construction) of scientific authority, see Shapin and Schaffer, *Leviathan and the Air Pump;* Latour, *Science in Action;* Knorr-Cetina, *Epistemic Cultures.*

CHAPTER 2. COGNITIVE MACHINERY AND EXPLANATORY AMBITIONS

1. This brief definition will serve us here. I return to the concept of naturalism in the last chapter, where I consider the significance of the wide variability, historical and other, of what we do think of as "nature."

2. See Saler, *Conceptualising Religion;* Dubuisson, *The Western Construction of Religion.*

3. Eliade, *Patterns in Comparative Religion*, xiii.

4. Gold, *Aesthetics and Analysis in Writing on Religion.* See also Segal, *Explaining and Interpreting Religion.*

5. Wiebe, *The Politics of Religious Studies*, ix. For comparable invocations and charges, see McCutcheon, *Manufacturing Religion* and *The Discipline of Religion;* Fitzgerald, *The Ideology of Religious Studies.*

6. Griffiths, "The Very Idea of Religion," 34. Earlier, to illustrate what he thinks ought to be meant by religion, Griffiths (himself Catholic) quotes—amusedly but sympathetically—Parson Thwackum's insistence: "When I mention religion, I mean the Christian religion; and not only the Christian religion, but the Protestant religion; and not only the Protestant religion but the Church of England" (from Henry Fielding, *Tom Jones*, quoted in Griffiths, 30–31).

7. Griffiths, "The Very Idea of Religion," 35. He derives the identification from Church history, etymology, and also, with unabashed perversity, from a suggestion by a scholar of Asian cultures, Timothy Fitzgerald, who urges that the term "religion" not be

used to describe non-European beliefs because it is overloaded with Christian connotations (*Ideology of Religious Studies*, 3–10).

8. See, e.g., Asad, *Genealogies of Religion*, 27–54; J. Z. Smith, "Religion, Religions, Religious"; Saler, *Conceptualising Religion;* Masuzawa, *The Invention of World Religions.*

9. Pyysiäinen, *How Religion Works;* Hinde, *Why Gods Persist.* Other key works include Lawson and McCauley, *Rethinking Religion;* Boyer, *Religion Explained;* and Atran, *In Gods We Trust.* Dennett, *Breaking the Spell*, synthesizes and promotes a number of these and related works.

10. Lawson and McCauley, *Rethinking Religion*, 1.

11. Since the 1990s, the project identified here as the New Naturalism has grown in size and become increasingly unified and institutionally established. Under such names as "cognitive studies of religion" or "the cognitive science of religion," it features the usual signs of an emerging academic field, including new graduate programs, specialized journals, international conferences, websites, and sizeable professional organizations (see esp. the International Association for the Cognitive Science of Religion [http://www.iacsr.com/Home.html]).

12. For an array of current charges or invocations of "scientism" in regard to the study of religion, see H. Smith, *Why Religion Matters*, and Stenmark, *Scientism.* For an attempt to distinguish among such charges or invocations, see Peterson, "Demarcation and the Scientistic Fallacy."

13. Wilson's views are echoed by promoters of a dubiously framed "conceptual integration" of the natural and social sciences (see, e.g., Tooby and Cosmides, "The Psychological Foundations of Culture") or, more recently, of a "vertical integration" of all the disciplines (see, e.g., Slingerland, *What Science Offers the Humanities*). I discuss the questionable assumptions supporting such views and projects in *Scandalous Knowledge*, 123–26.

14. In *Rethinking Religion*, the New Naturalists Lawson and McCauley themselves deplore "scientism," but in a rather ad hoc sense. Defining it as a "logical positivist" confinement of explana-

tions in religious studies to the types of causal theories found in the physical sciences, they frame it as the counterpart to an exclusivist "humanism" that would confine the field to interpretive studies and go on to recommend a "Third Way" that recognizes explanation and interpretation as "interactive" and "complementary." Although much of what Lawson and McCauley write along these lines is perceptive and useful, their typology seems designed primarily to open a space for the theory of religious ritual they develop in the book. Such a Third Way is needed, it appears, because their theory, modeled closely on Noam Chomsky's transformational-generative linguistics, is clearly not what most scholars would call interpretive but, especially for those (by no means only logical positivists) who question Chomskian assumptions and methods, not very explanatory either.

15. Boyer's title echoes Dennett, *Consciousness Explained*, and evokes the similarly chest-thumping mode of other recent titles in related fields, including Pinker, *How the Mind Works*, echoed by Pyysiäinen, *How Religion Works.*

16. Boyer, *Religion Explained*, 2, 4–6.

17. Ibid., 16–17, 21, 50.

18. I repeat here but do not endorse the purposive idiom ("evolved 'to' provide") of evolutionary psychology. See the next section for related discussion.

19. For important challenges to the computational-modular model, see Port and van Gelder, eds., *Mind as Motion;* Hendriks-Jansen, *Catching Ourselves;* Woodward and Cowie, "The Mind Is Not (Just) a System of Modules Shaped (Just) by Natural Selection." For important alternative models of cognition, see Núñez and Freeman, eds., *Reclaiming Cognition;* Melser, *The Act of Thinking;* Hutchins, *Cognition in the Wild;* Sterelny, *Thought in a Hostile World;* Pfeifer and Bongard, *How the Body Shapes the Way We Think;* Rockwell, *Neither Brain nor Ghost;* Wheeler, *Reconstructing the Cognitive World.*

20. On the significance of experiential learning, Woodward and Cowie observe: "[The] view that the modules existing in the

adult mind are largely genetically specified . . . is inconsistent with what is known about the role of experience-dependent learning and development in shaping the mature mind. . . . Less modular structures and capacities that are present in infants interact with both the environment and genetic mechanisms to generate new competences that were not directly selected for" ("The Mind Is Not," 313–14). On the interactive dynamics involved in the transmission of skills and beliefs, see Hendriks-Jansen, *Catching Ourselves;* on the crucial operation of cognitive resources, see Hutchins, *Cognition in the Wild,* and Sterelny, *Thought in a Hostile World.*

21. Boyer also adopts from evolutionary psychology the idea that a proper causal explanation of various features of human behavior and culture—from concepts to social practices—can be arrived at through what is called "functional analysis": "More generally, social institutions are around and people comply with them, because they serve some *function.* Concepts too have functions and that is why we have them. If you can identify the function, you have the explanation" (*Religion Explained,* 35). As distinct from the social functions identified by theorists in the "social-functionalist" approach (Marx, Durkheim, Weber, and so forth) that Boyer and other evolutionary psychologists reject, the functions identified in evolutionary psychology are those that presumably contribute to the reproductive fitness of individuals.

22. For the history of such views, see Bowler and Morus, *Making Modern Science,* esp. 4–9. For a summary of recent views of explanation, see Keil and Wilson, "Explaining Explanation."

23. See Boyer, *Religion Explained,* 48. To support their conjectures, Boyer and other evolutionary psychologists adduce empirical studies (for example, ethnographic field reports, laboratory experiments with human subjects, or brain-imaging studies) and reasoning of various kinds, including, as noted above, functional analysis or so-called reverse engineering. But the empirical evidence is sometimes meager or dubious at its source, the reasoning is often strained, and the strictly hypothetical nature of the mechanisms that are posited tends to get lost in the elaboration of the explanations. For illus-

tration and discussion of these points with regard to evolutionary psychology, see Woodward and Cowie, "The Mind Is Not (Just)"; B. H. Smith, *Scandalous Knowledge*, 130–52; R. C. Richardson, *Evolutionary Psychology as Maladapted Psychology;* and chapter 3 below.

24. See Barrett, *Why Would Anyone Believe in God?*, for another extended description of the posited mechanism, referred to as HADD (hyperactive agent-detection device).

25. While such mechanisms or tendencies may help explain how ideas of supernatural beings originated in some postulated prehistoric era, they do not of course explain how people come to have such ideas when—as has commonly been the case throughout recorded human history—they are born into cultures (including practices, institutions, and artifacts) that are saturated with the supernatural and often with specifically theistic allusions and images. To explain this sort of origination of religious belief, Boyer and other New Naturalists typically invoke culturally contagious "memes" and comparably dubious "epidemiological" mechanisms of strictly mind-to-mind transmission.

26. Hume, *The Natural History of Religion*, 141, punctuation slightly edited. The first part of the passage is quoted by Atran (*In Gods We Trust*, 68) and Dennett (*Breaking the Spell*, 108); it also suggested the title of Guthrie, *Faces in the Clouds.*

27. Nietzsche, *Genealogy of Morals*, 45.

28. Boyer, *Religion Explained*, 257. The tendency could also be illustrated by the purposive idiom of evolutionary psychology noted above, in which the quasi-agent Natural Selection is said to have "engineered" traits "to" perform particular adaptive "functions."

29. Tylor, *Religion in Primitive Culture;* Frazer, *The Golden Bough;* Gomperz, *Greek Thinkers.* Gomperz's anticipations include the idea that many features of archaic religions are products of universal impulses and instincts that survive in civilized man, that the psychological tendencies in question are embodied in the nervous system, and that concepts of the supernatural are the result of patterns of inference-drawing—based, he ventured, on dreams, self-observation, and observation of such phenomena as shadows, re-

flections, perfumes, the wind, and the effects of death on the bodies of animals and humans (*Greek Thinkers*, I., 15–24). Gomperz also notes that the resulting ideas often contradict the official dogmas of the relevant creeds, here anticipating a point elaborated in Slone, *Theological Incorrectness*.

30. Boyer does not mention Hume, Nietzsche, Weber, Freud, or Durkheim and dismisses *The Golden Bough* as "a sterile compilation" (*Religion Explained*, 55). See Dennett, *Breaking the Spell*, for the exaggerated claim, discussed in chapter 5 below, that there has been a "taboo" on the naturalistic explanation of religion until very recently. See R. Smith, *Being Human*, for pertinent comments on the ahistorical character of sociobiology and related evolutionary accounts of human behavior.

31. See Boyer, *Religion Explained*, 6.

32. Like Nietzsche's *Genealogy of Morals*, Hume's *Natural History of Religion* is an account of ideas and practices as products of dynamic processes, both psychological and social. See especially Hume's analysis of what he posits as a cycle of monotheistic and polytheistic beliefs (*Natural History of Religion*, 158–60).

33. Boyer, *Religion Explained*, 226–28.

34. Not all New Naturalists reject such explanations of ideas of immortality. See especially, as discussed in chapter 3 below, Atran, *In Gods We Trust*, 269–70.

35. Boyer, *Religion Explained*, 122–29. Boyer and many other New Naturalists reject group-level selection as technically implausible (see also Atran, *In Gods We Trust*, 227–34) and, accordingly, also reject the group-selectionist account of morality (and other features of religion) put forth in D. S. Wilson, *Darwin's Cathedral*.

36. Boyer, *Religion Explained*, 174–91. The quoted passages occur on 186–87.

37. Ibid., 191, 202.

38. Some recent studies allied to the New Naturalism are more broadly interdisciplinary and also more responsive to other views of cognition (see, e. g., Schüler, "Embodied Cognition"), but narrowly defined "cognitive" and "Darwinian" approaches to the study of reli-

gion continue to be promoted. The spectrum of current approaches is usefully represented by the contributions to Bulbulia et al., eds., *The Evolution of Religion.*

39. Boyer, in accord with the typology of explanations described above, would presumably classify these general observations with what he calls, in *Religion Explained,* "intellectualist" accounts of religion, and, since Burkert does not offer "causal mechanisms" to explain the experiential effects, would also presumably reject the observations as inadequate.

40. Burkert, *Creation of the Sacred,* 81, 83, 81–85.

41. Ibid., 82 (italics added), 97, 95–96.

42. See *Creation of the Sacred,* 8–23, for Burkert's sympathetic but distanced assessment of E. O. Wilson's early program of sociobiology.

43. The book has been generally well received by educated nonspecialist readers. Its reception by variously specialized readers follows largely disciplinary lines and related goals, expectations, and measures of success. Thus scholar of religious studies Tomoko Masuzawa sees *Creation of the Sacred* as improperly biologizing the field (Masuzawa, "In Nature's Trail") while cognitive and evolutionary theorist Daniel Dennett, though acknowledging the book's achievements, expresses dismay at what he sees as Burkert's concessions to "the hair-trigger sensitivities of his fellow humanists" (Dennett, *Breaking the Spell,* 264). The objections of Catholic theologian John Haught to what he sees as Burkert's shallow Darwinian reductionism (Haught, *Deeper than Darwin*) are discussed in chapter 4 below.

44. Boyer, *Religion Explained,* 232–33.

45. Of course participants' characteristic experiences of a ritual are not the same as their explicit interpretations of its meaning. If, however, one is attempting to explain what secures a particular ritual's cultural reproduction and survival, such interpretations might be thought relevant data.

46. For a classic theorization, see Durkheim, *The Elementary Forms of Religious Life,* 303–412. For a comprehensive review, see Bell, *Ritual.*

47. Boyer, *Religion Explained*, 262, 257–63. Boyer follows Lawson and McCauley's highly speculative theory of rituals, which he describes as "a very precise account" (259).

CHAPTER 3. "THE GODS SEEM HERE TO STAY"

1. Atran, *In Gods We Trust*, 4–5.
2. Ibid., 50. Atran's term "function" here reflects the idea that, to survive, any persistent behavioral or cultural phenomenon must pay its way in biological-fitness coin. For discussion of the idea and some of its problems, see R. C. Richardson, *Evolutionary Psychology as Maladapted Psychology*, 41–88.
3. Atran, *In Gods We Trust*, 280.
4. See ibid., 50.
5. Ibid., 56–57.
6. Ibid., 88.
7. Atran's classification of certain beliefs as "symbolic" as distinguished from "commonsense" is discussed below.
8. Atran, *In Gods We Trust*, 211.
9. As noted in chapter 2 above, Chomsky's argument and theory are central to the "cognitive" account of religious rituals in Lawson and McCauley, *Rethinking Religion*. In the view of various critics of the argument, mental-computational competencies, mechanisms, or devices with the physiological and functional specifications required by the theories at issue are inconsistent with human brain anatomy and with current empirical work in genetics and paleoanthropology. For important critical discussions of the Language Acquisition Device proposed by Chomsky in the 1960s, see Deacon, *The Symbolic Species;* Cowie, *What's Within?;* Lieberman, *Toward an Evolutionary Biology of Language.*
10. Atran, *In Gods We Trust*, 71–78. Atran makes much of an experimental finding to the effect that children do not associate God with their parents. The generality of the particular finding is dubious and its weight in relation to the validity of Attachment Theory certainly slighter than Atran suggests.
11. Ibid., 65. In Atran's purely "cognitive" account, the bun-

dling of such features, like virtually everything else involved in the generation of concepts of gods, reflects only the operation of innate mental mechanisms.

12. For the casting of "social constructionism" and/or empiricist models of mind as heresies or exploded myths, see Tooby and Cosmides, "The Psychological Foundations of Culture," and Pinker, *The Blank Slate.* For anxieties about cultural relativism, see Atran, *In Gods We Trust*, 56–57, 84–88, 143–44.

13. See Núñez and Freeman, *Reclaiming Cognition*, for relevant work in neuroscience. For the epistemological issues, see B. H. Smith, *Scandalous Knowledge.* The conceptual and historical relations among rationalism, mentalism, and nativism, and their joint opposition to empiricism, are explored in Cowie, *What's Within?*

14. See the numerous invocations of the Enlightenment by E. O. Wilson in *Consilience* and by Dennett and other promoters of cognitive-evolutionary approaches to religion in the online forum, *Beyond Belief: Enlightenment 2.0.*

15. According to this theory, one of the main reasons it is easy for theistic concepts to be accepted, transmitted, and remembered is that gods and other supernatural beings (spirits, ancestors, and so forth), in being attributed with the ability to see us and hear our praises, conform to our innate intuitions about what personal agents can do. But (the theory continues), in having such attributes as invisibility, omnipotence, and immortality, such beings violate our innate intuitions about the properties of persons, and this is what makes ideas and stories about them especially striking and memorable. See Atran, *In Gods We Trust*, 83–113. For other invocations of the theory, see Boyer, *Religion Explained*, 51–91; Pyysiäinen, *How Religion Works*, 18–23, 217–34; Dennett, *Breaking the Spell*, 116–51.

16. Atran, *In Gods We Trust*, 86, 87.

17. d'Aquili and Newberg, "The Neuropsychological Basis of Religions."

18. Atran, *In Gods We Trust*, 183.

19. Ibid., 143–44.

20. Bell, "The Chinese Believe in Spirits," 107, 109, 111; Veyne,

Did the Greeks Believe in Their Myths?, xi. Veyne comments that, having set out "to study the plurality of the modalities of belief—belief based on word, on experience, and so on," he came in the course of his studies "to recognize that, instead of speaking of beliefs, one must actually speak of truths, and that these truths were themselves products of the imagination" (ibid.).

21. Orsi, *Between Heaven and Earth*, 203. On the heterogeneity of contemporary beliefs, both declaredly religious and ostensibly secular, see also Hervieu-Léger, "In Search of Certainties," 59–69.

22. Dennett discusses the work of, among others, Boyer, Atran, and Walter Burkert, the last rather condescendingly ("[Burkert] has begun educating himself in the evolutionary biology that he sees clearly must ground his own efforts at theorizing" [*Breaking the Spell*, 263–64]). While Dennett allows that not all religious ideas are toxic, he does not say either how the status of a particular idea would be determined or what authorities would enforce the recommended quarantines or direct the efforts at conviction-eradication.

23. Dennett, *Breaking the Spell*, 221–22, 227–28. Because mental states about supernatural entities do not meet the specifications for the proper use of the term "belief" arrived at by formal philosophical analysis, Dennett calls religious belief a "misnomer" (ibid., 318) and suggests that it be renamed something else like religious *conviction* or religious *professing* (see the passages quoted below). For discussion of the dubious determinative authority of such "conceptual analysis" both generally and with particular regard to the term and/or concept *belief*, see B. H. Smith, *Belief and Resistance*, 96–100.

24. Dennett, *Breaking the Spell*, 228, 233. He observes that the creed of the Eucharist is "cleverly shielded from . . . concrete tests" of its "truth" (that is, whether the wine really turns into blood) and quotes Richard Dawkins: "The meme for blind faith secures its own perpetuation by the simple unconscious expedient of discouraging rational inquiry" (ibid., 227–28, 230; quoting Dawkins, *The Selfish Gene*, 212–13).

25. For recent detailed studies and discussion, see Orsi, *Between Heaven and Earth*, and Mahmood, *The Politics of Piety*. For a

subtle analysis of the significance of embodiment in religious observance, see Keane, "The Evidence of the Senses and the Materiality of Religion."

26. Dennett, *Breaking the Spell*, 210–16, 222.

27. Ibid., 318.

28. For a sociologically and historically informed review of the complex relations between the secular and the religious in contemporary societies, see Casanova, "Rethinking Secularization." As Casanova remarks, "A proper rethinking of secularization will require a critical examination of the diverse patterns of differentiation and fusion of the religious and the secular and their mutual constitution across all world religions" (10).

29. For the former, see, e.g., Dennett, *Breaking the Spell*, as discussed above. For the latter, see, e.g., Boyer, *Religion Explained;* McCauley, "The Naturalness of Religion." McCauley's views in this matter are discussed in chapter 5 below.

30. Atran, *In Gods We Trust*, 280.

31. Ibid., 66–67.

32. Boyer, *Religion Explained*, 226–28. See the discussion in chapter 2 above.

33. Atran, *In Gods We Trust*, 117, 278, 144–45 (italics in original text).

34. Atran intends the analysis to illustrate how an otherwise severely limited (because it is "mindblind") anthropology is improved by the introduction of "cognitive models" (ibid., 219).

35. Diamond, *Collapse*.

36. Atran, *In Gods We Trust*, 223–27.

37. Ibid., 279–80, 210.

38. Ibid., 57. See also Dennett, *Breaking the Spell*, where, responding to the observation by a contemporary believer that God need not be thought of as a supernatural entity, Dennett writes with evident exasperation: "If God is not some kind of supernatural entity, then *who knows* whether you or I believe in him (it?)?" (216).

39. Boyer, *Religion Explained*, 50.

40. Boyer does not seem to mean this co-activation of cog-

nitive mechanisms as a formal definition of religion. He does not offer an explicit definition of "religion" anywhere in *Religion Explained* but, like a number of other New Naturalists, uses the term as equivalent to beliefs and/or practices involving posited supernatural agents.

41. Atran implicitly recognizes these possibilities in a section of his book titled "Why Are Mickey Mouse and Marx Different from God?," but represents the questions they raise only as limits of certain cognitive theories of religion, to be supplemented accordingly by "commitment" theories (*In Gods We Trust*, 13–15).

42. Bloch, "Are Religious Beliefs Counter-Intuitive?," 136–37.

43. *In Gods We Trust*, 93.

44. Bloch, "Are Religious Beliefs Counter-Intuitive?," 145. Bloch mentions those, such as political leaders and police, who, by virtue of their cultural roles and status, have a certain "aura" and evoke special attitudes and behavior. One could readily add judges, teachers, sports figures, and media celebrities.

CHAPTER 4. DEEP READING

1. Other prominent examples include *Wrestling with the Divine* by biological theorist/theologian Christopher C. Knight; *Life's Solution* by paleontologist Simon Conway Morris; and *Finding Darwin's God* by biologist Kenneth Miller.

2. For good historical accounts, see Brooke and Cantor, *Reconstructing Nature;* Thomson, *Before Darwin.*

3. The quoted terms and phrases are found throughout Haught's *Deeper than Darwin* and in other works cited above.

4. A number of recent works by other theologically inclined scientists, including *The Mind of God* by physicist-mathematician Paul Davies and *The Language of God* by geneticist Francis Collins, are more in the mode of traditional natural theology. That is, they seek to demonstrate not merely that scientific knowledge and religious teachings can be reconciled but that various findings of modern science give positive evidence of the existence and attributes of a more or less traditional personal and providential deity.

5. Haught, *Deeper than Darwin*, 59.

6. The language tends to be more quasi-technical and the attempted conceptual synthesis more schematic when the effort is not merely to reconcile science and religion but, more ambitiously, to "integrate" or, as it is sometimes said, to "marry" them (see, e.g., *The Marriage of Sense and Soul* by New Age theorist Ken Wilber). The resulting style evokes various occult traditions (for example, Rosicrucianism) with which such integrative projects share many thematic preoccupations, sexual union and access to cosmic powers among them. On the rhetoric of natural theology and other compatibilist efforts, see Brooke and Cantor, *Reconstructing Nature.* On the rhetoric of theology more generally, see Burke, *The Rhetoric of Religion.*

7. Haught, *Deeper than Darwin*, 58.

8. Ibid., 60–62.

9. Ibid., 65.

10. For a good account of that double inhabitation, see Lipton, "Science and Religion." For a sophisticated example and theorization of it, see Latour, "Thou Shall Not Freeze-Frame," where, as elsewhere in his work (see, e.g., *Iconoclash*), Latour elaborates significantly nonstandard understandings of both "science" and "religion."

11. The intellectual resources of theological critiques of scientific explanations are inevitably limited, but, of course, theologians, like other informed commentators, can indicate conceptual problems in such explanations and note historically inaccurate claims made in support of them. See, e.g., Hart, "Daniel Dennett Hunts the Snark."

12. Tooby and Cosmides, "Psychological Foundations of Culture," 48.

13. Lawson and McCauley, *Rethinking Religion.* The association of spatial metaphors (superficial/deep, external/interior) with epistemological or metaphysical dualisms (illusory/true, contingent/essential) appears to be classic and chronic.

14. Boyer, *Religion Explained;* Pyysiäinen, *How Religion Works.*

15. Haught, *Deeper than Darwin*, 104, 106.

16. For some Darwinians, notably David Sloan Wilson and other scholars proposing group-selectionist accounts of religion, it *would* be the rituals, doctrines, etc., that were adaptive (see D. S. Wilson, *Darwin's Cathedral*). But the New Naturalists generally, Boyer, Atran, and Dawkins among them, explicitly reject the view that religious teachings or practices, as such, are adaptive and are highly critical of the idea of group-level selection. Haught does not, in any case, appear concerned with that dispute here.

17. Haught, *Deeper than Darwin*, 109.

18. See, for example, Bloom, "Is God an Accident?," where the answer given to the title question is affirmative.

19. See Dawkins, *The Selfish Gene;* Dennett, *Darwin's Dangerous Idea.*

20. Of course, various aspects of neo-Darwinism may be rejected on other empirical or conceptual grounds. The nature of genes and their operation in individual development are matters of ongoing research and dispute in theoretical biology and evolutionary theory (see, e.g., Lewontin, *The Triple Helix;* Keller, *The Century of the Gene;* Oyama, *The Ontogeny of Information;* Richerson and Boyd, *Not by Genes Alone*). So is the exact role of natural selection in the evolution of species (see, e.g., Goodwin, *How the Leopard Changed Its Spots;* Weber and Depew, "Developmental Systems, Darwinian Evolution, and the Unity of Science").

21. Haught, *Deeper than Darwin*, 106.

22. Haught's characterization also recalls Atran's biological economics as discussed in chapter 3 above, but he does not cite Atran's work in *Deeper than Darwin.*

23. Burkert, *Creation of the Sacred*, 84, 128, 154.

24. Haught, *Deeper than Darwin*, 109, 110. Haught is not alluding here to Daniel Dennett or Richard Dawkins, whose "new atheist" books, *Breaking the Spell* and *The God Delusion*, appeared after *Deeper than Darwin.* These two evolutionists (unlike Burkert, Boyer, or, later, Atran) *are* centrally and determinedly "critics of religion" and, indeed, disturbed by the "wallowing" in error of most humans

in just the old-fashioned Enlightenment way that Haught represents as more consistent and moral. As for what Haught sees as the atheism of Darwinians in religious studies ("apparently the new Darwinian debunkers of religion" have either "give[n] up" "religious ideas" or "never had" them), it may be noted that while not all New Naturalists declare their personal religious views explicitly, a number of them, like many other scholars pursuing naturalistic approaches to religion in the past (for example, anthropologists Victor Turner and Mary Douglas), are themselves religiously identified (see, e.g., among the New Naturalists, Barrett, *Why Would Anyone Believe in God?).*

25. For the positive response, see Bell, "Modernism and Postmodernism in the Study of Religion"; for the negative response, see Segal, "All Generalizations Are Bad"; Dennett, *Breaking the Spell*, 262–63, 370–72.

26. See Dennett, *Breaking the Spell*, 200–46, discussed in chapter 3 above.

27. Haught, *Deeper than Darwin*, 200n22.

28. Examples include Dennett, *Breaking the Spell*, 312–13, and Wolpert, *The Unnatural Nature of Science*, both discussed in chapter 5 below.

29. For related discussion of the emergence of contemporary science studies and its outraged, scornful reception by various philosophers and scientists, see B. H. Smith, *Scandalous Knowledge*, 46–107. Sociologist David Bloor compares the hostile reception of the "strong program" in the study of scientific knowledge to that which greeted the Tübingen School's comparably radical historical studies of religion (*Knowledge and Social Imagery*, 184–85). For a purported demonstration that constructivist (or, as he calls them, "relativist") views of knowledge are self-refuting, see Paul Boghossian, *Fear of Knowledge*. For a comment on Boghossian's characteristic misrepresentations and strained arguments in this regard, see B. H. Smith, "Reply to an Analytic Philosopher."

30. Griffiths, "Faith Seeking Explanation," 56. For a similar series of moves by Griffiths against naturalism-promoting scholar

of religion Russell T. McCutcheon four years earlier, see his "*Tu Quoque*," discussed in chapter 1 above. For an analysis of the general logical-rhetorical structure and typical intellectual occasions of charges of self-refutation and/or "performative" contradiction, see B. H. Smith, *Belief and Resistance*, 73–104.

31. Boyer, *Religion Explained*, 321.

CHAPTER 5. REFLECTIONS

1. For useful recent discussions, see Nielsen, *Naturalism and Religion*, and Flanagan, "Varieties of Naturalism."

2. Griffiths, "Very Idea of Religion," esp. 34–35.

3. See, e.g., cosmologist Vilenkin, *Many Worlds in One*, for the idea of the existence (in some sense) of multiple possible worlds; see Smolin, *The Trouble with Physics*, for the view that longstanding foundational—and, in effect, metaphysical—assumptions of physics are hobbling the development of more conceptually adequate and empirically responsive theories in the field.

4. For a comparable observation, see Nagel, "The Fear of Religion." For Nagel, the understandable (he says "natural") impulse to extend "as far as it will go" the successful empirical method of the natural sciences has "gotten out of control." He observes that the resulting hypertrophic impulse, which he refers to as "reductive physicalism," turns the descriptions of the world given by the natural sciences into "an exclusive ontology" and "denies reality to what cannot be so reduced" (28). While I do not share all Nagel's anxieties here and would not make all his arguments, I think he is correct to observe this crucial slippage between methodological and metaphysical naturalism in the work of Dawkins and others involved in the current science-invoking polemics against "God" or "religion."

5. See, e.g., R. Williams, "Nature," and Hadot, *The Veil of Isis*.

6. The specific lexeme "supernatural" seems to have had a late medieval origin. Aquinas used the term *supernaturalis* in the thirteenth century. The *Oxford English Dictionary* gives 1526 as its earliest recorded usage in English. For extensive field research and

discussion of current differences of usage and conception in these regards, see Descola and Pálsson, eds., *Nature and Society.*

7. For such charges, see Segal, "All Generalizations Are Bad."

8. Some of us may attempt to delineate our personal constructions of the universe to friends or associates, but, precisely, as *ours.* This is significantly different from claiming that we have discovered, deduced, or intuited some way that the universe objectively *is.*

9. For example, Daniel Dennett writes, presumably with some self-mockery (but maybe not): "If I found that I believed in poltergeists or the Loch Ness Monster, I'd be, well, embarrassed. . . . I might take steps to cure myself of this awkward bulge in my otherwise impeccably hardheaded and rational ontology" (*Breaking the Spell,* 221).

10. Boyer, *Religion Explained,* 320. Boyer's suggestion that there was, in the West, a specific, independent realm of "empirical statements" into which Church authorities could be said to have "made the . . . mistake of meddling" has no historical basis. See the discussion and references below for the fundamental institutional and intellectual connections between the Church and science for a large part of the history of both. For an account of the emergence of the institutions and practices of empirical science in the seventeenth century, see Shapin and Schaffer, *Leviathan and the Air Pump.*

11. Boyer, *Religion Explained,* 321–22. Boyer writes: "As philosopher Robert McCauley concludes, on the basis of similar arguments, science is every bit as 'unnatural' to the human mind as religion is 'natural.'" He also cites Wolpert in this connection. The mutual citations among Boyer, McCauley, and Wolpert and of each of them by other New Naturalists (e.g., Jason Slone, *Theological Incorrectness*) are frequent.

12. McCauley, "The Naturalness of Religion and Unnaturalness of Science," 66, 77, 80. In *Rethinking Religion,* the book McCauley co-authored with E. Thomas Lawson in 1990, the historical and intellectual connections between religion and science are clearly noted. Moreover, in its influential first chapter, *Rethinking Religion*

develops a strong, carefully argued rejection of logical-empiricist conceptions of scientific method, explanation, and progress in favor of distinctly post-Kuhnian conceptions. The more recent essay by McCauley under discussion here seems to represent a subsequent change of position for him.

13. On the institutional and ideological continuities between the medieval church and modern science, see Noble, *World without Women*. For the significance of Newton's theological interests, see Janiak, "Introduction" to Newton, *Philosophical Writings*, and *Newton as Philosopher*. For an analysis of the historical relations between the two terms at issue, see Harrison, "'Science' and 'Religion.'" For a general account respectful of the complexities of the interactions between the two, see Olson, *Science and Religion, 1450–1900*.

14. For the prevalence of religious ideals and themes in eighteenth- and nineteenth-century literature, philosophy, and science, see Abrams, *Natural Supernaturalism*. Kirschner, "Sources of Redemption in Psychoanalytic Developmental Psychology," discusses the significance of secularized biblical accounts of historical design for developmental psychology; Kessen, "The Transcendental Alarm," examines salvationist themes in Social Darwinism. On the appropriation of science as both a repudiation of and substitute for religion in the United States, see Hollinger, *Science, Jews, and Secular Culture*. For examples of religious teleological, meliorist narrative, see Haught, *Deeper than Darwin*, 55-68, as discussed in chapter 4 above; for secular examples, see the editorial page in virtually any issue of the journal *Science*. On patriarchal norms and/or misogyny in Christianity and Western science, see Noble, *World without Women*.

15. See, e.g., *Religion Explained*, 311, where Boyer writes: "The processes that create 'belief' are the same in religion and in everyday matters." This was also the emphasis in his earlier book, *The Naturalness of Religious Ideas*.

16. McCauley, "The Naturalness of Religion," 66–67. As an example of the generality of cognitive biases, he mentions confirmation bias, the occurrence of which among scientists is well estab-

lished (see Mynatt, Doherty, and Tweney, "Confirmation Bias"). See also de Cruz and de Smedt, "The Role of Intuitive Ontologies in Scientific Understanding," for the idea that various once-adaptive intuitions that "remain an integral and stable part of human cognition" also "continue to play a role in scientific understanding" (365); they cite studies on essentialist thinking about species by, among others, Atran and Boyer.

17. For early influential accounts of belief-persistence or cognitive conservatism among scientists, see Fleck, *Genesis and Development;* Kuhn, *The Structure of Scientific Revolutions.* The sharing by groups of scientists of socially conservative views has been extensively documented by cultural and intellectual historians; see, e.g., Proctor, *Value-Free Science?*

18. McCauley, "The Naturalness of Religion," 68, 71.

19. For the extensive historical connections between science and technology, see, e.g., Bowler and Morus, *Making Modern Science,* esp. 391–414; Cook, *Matters of Exchange.* For detailed description and extensive analysis of related developments in twentieth-century science, see Shapin, *The Scientific Life.* Shapin also documents the continued and increasingly dubious effort on the part of some scientists and academic commentators to distinguish sharply between, on the one hand, "pure" or "basic" science and, on the other, technology or "applied" science.

20. McCauley, "The Naturalness of Religion," 82.

21. Recent studies by sociologists of religion indicate the continuation of secularizing trends in much of the United States as well as Western Europe, with traditional faiths being increasingly replaced by various amalgams of secular and religious belief (see, e.g., Davie, *Religion in Britain since 1945* and *Religion in Modern Europe*). Also, as noted in chapter 4, the trend in Western theology is science-compatibilism, not the questioning of scientific knowledge as such, much less the effort to extinguish scientific activity.

22. Gopnik, "Explanation as Orgasm and the Drive for Causal Knowledge," 300, 301. McCauley's essay appears in the same collection as Gopnik's.

23. See B. H. Smith, *Scandalous Knowledge*, 108–29. The ideology in question is not the observation of extensive and perhaps fundamental differences between the sciences and humanities—in their domains of inquiry, methods, aims, achievements, intellectual styles, and so forth—or even the observation that they operate in some ways as two distinct "cultures." It is the perception and representation of those differences in polarized, hierarchical, and invidious terms: as, for example, important versus trivial research, deep versus superficial knowledge, or the familiar "hard" versus "soft" disciplines.

24. Dennett, *Breaking the Spell*, 188, italics in original. The entire passage is in parentheses in the original text.

25. Throughout *Breaking the Spell*, Dennett refers to scientists individually by name, summarizes their views conscientiously, and characterizes those views, even when he disagrees with them, with respect. Humanists are routinely lumped together and characterized disdainfully or, if referred to individually and approvingly, as in the case of Walter Burkert, then with marked condescension (see, e.g., ibid., 263–64).

26. For a classic and still compelling set of arguments, see Smart, *The Science of Religion and the Sociology of Knowledge.*

27. For a defense of the value of participants' perspectives in connection with specifically "cognitive" approaches to the study of religion, see Keane, "The Evidence of the Senses."

28. In musicology, see, e.g., from the late 1950s, Meyer, *Emotion and Meaning in Music;* in art history, see Arnheim, *Visual Thinking;* Gombrich, *Art and Illusion;* Baxandall, *Painting and Experience in Fifteenth Century Italy;* Alpers, *The Art of Describing.*

29. See, e.g., Spilka et al., eds., *The Psychology of Religion*, the bibliography of which lists over three thousand titles dating from the late nineteenth century. Other important naturalistic accounts of religion, e.g., by Hume, Tylor, Gomperz, Durkheim, and Weber, are cited and discussed in chapter 2 above.

30. This includes Clifford Geertz, whose work is often invoked as exemplifying an alleged anti-evolutionary or anti-biological bias

in interpretive approaches to the study of culture. See esp. "Religion as a Cultural System," in Geertz, *Interpretation of Cultures*, 87–125.

31. For comparably melodramatic and tendentious accounts of intellectual history, see Tooby and Cosmides, "Psychological Foundations," 21; E. O. Wilson, *Consilience*, 14–44; Pinker, *The Blank Slate*, 14–58. I discuss them in *Scandalous Knowledge*, 145–46.

32. For such claims, see, e.g., J. Carroll, *Literary Darwinism*. To determine the meaning and explain the appeal of novels such as Jane Austen's *Pride and Prejudice*, Carroll applies to their authors, readers, and texts what he believes he has learned from evolutionary psychology about the fundamental, universal traits of human nature, mammalian mating-strategies among them. I reserve discussion of such claims, methods, and objectives for a future occasion.

33. The limited scientific education of students and scholars working in the humanities, like the limited education in the humanities of students and researchers working in the natural sciences, can be seen as both a product of the dominant Two Cultures ideology and a source of its perpetuation. Other factors, however, are involved. These include institutional pressures on faculty and students in the sciences related to ever-increasing specialization and pre-professionalism and, in both the sciences and the humanities, curricular or pedagogic traditions that discourage Two Cultures crossings in either direction. For discussion, see B. H. Smith, *Scandalous Knowledge*, 108–15.

34. The key words here are *indiscriminate* for the endorsements and *crude*, *glib*, *pedestrian*, and *inert* for the evolution-invoking importations. In literary studies as in other humanities-based fields, scientifically informed scholars may incorporate concepts and findings from evolutionary biology, cognitive science, and other natural-science fields in ways that can be subtle, original, illuminating, and sometimes significantly transformative for their own fields. As indicated in chapter 2 above, an important example of the possibility for the field of religious studies is Burkert, *Creation of the Sacred*. It is relevant here, however, that Burkert does not make claims of scientific status for his own contributions to the field; that he retains the

aims, idioms, and perspectives of humanities-based scholarship even while deploying concepts and findings from evolutionary biology and other natural sciences; and that, while he is respectful of the program of sociobiology, his invocations of it are not uncritical.

35. For minimally informed dismissive identifications of recent revisionary accounts of scientific knowledge with "postmodernism," see E. O. Wilson, *Consilience*, 42, 182; Slingerland, *What Science Offers*, 74–98; Dennett, *Breaking the Spell*, 262–63. Later in *Breaking the Spell* Dennett invokes, without citation, the "clueless," "misguided work" of the "early days" of science studies, now supposedly "more than balanced" by the "recent success" of "deeply informed and comprehending work" by practitioners (313). As anyone in the fields involved would agree, however, the founders and early practitioners of what we now call science studies were as deeply informed about the subject or subjects of their research as current practitioners. And, as intellectual historians could readily document, it was precisely from the perspective of outraged scientists and traditionalist philosophers of science that the works of such early practitioners as Kuhn, Bloor, Foucault, Pickering, or Latour were judged "clueless" or "misguided." The inaccuracy of Dennett's account here as well as the vagueness of his allusions and the equivocality of his assessments (e.g., he writes, again without citations, of "all the misuse" to which Kuhn's "wonderful book" has been put [404n4]) suggest that his characterizations of science studies are largely second-hand, drawn, it appears, from hostile reports by just such outraged scientists and traditionalist philosophers.

36. Nietzsche, *Genealogy of Morals*, 151, 155, italics in original.

37. Of course, neither shamanism nor Presbyterianism is "just a belief" either.

Works Cited

Abrams, M. H. *Natural Supernaturalism: Tradition and Revolution in Romantic Literature*. New York: Norton, 1971.

Alpers, Svetlana. *The Art of Describing: Dutch Art in the Seventeenth Century*. Chicago: University of Chicago Press, 1983.

Arnheim, Rudolf. *Visual Thinking*. Berkeley: University of California Press, 1969.

Asad, Talal. *Genealogies of Religion: Discipline and Reasons of Power in Christianity and Islam*. Baltimore: Johns Hopkins University Press, 1993.

Atran, Scott. *In Gods We Trust: The Evolutionary Landscape of Religion*. Oxford: Oxford University Press, 2002.

Bader, Chris. "When Prophecy Passes Unnoticed: New Perspectives on Failed Prophecy." *Journal for the Scientific Study of Religion* 38, no. 1 (1999): 119–31.

Bainbridge, William Sims. *The Sociology of Religious Movements*. New York: Routledge, 1997.

Barkow, Jerome, Leda Cosmides, and John Tooby, eds. *The Adapted Mind: Evolutionary Psychology and the Generation of Culture*. New York: Oxford University Press, 1995.

Barrett, Justin L. *Why Would Anyone Believe in God?* Walnut Creek, CA: AltaMira Press, 2004.

Baxandall, Michael. *Painting and Experience in Fifteenth Century Italy: A Primer in the Social History of Pictorial Style.* Oxford: Clarendon Press, 1972.

Beckford, James A. *Social Theory and Religion.* Cambridge: Cambridge University Press, 2003.

Bell, Catherine M. "'The Chinese Believe in Spirits': Belief and Believing in the Study of Religion." In *Radical Interpretation in Religion,* edited by Nancy K. Frankenberry, 100–16. Cambridge: Cambridge University Press, 2002.

———. "Modernism and Postmodernism in the Study of Religion." Review of *Explaining Religion* by J. Samuel Preus, *Rethinking Religion* by E. Thomas Lawson and Robert N. McCauley, and *Genealogies of Religion,* by Talal Asad. *Religious Studies Review* 22, no. 3 (1996): 179–90.

———. *Ritual: Perspectives and Dimensions.* New York: Oxford University Press, 1997.

Berger, Peter. *The Sacred Canopy: Elements of a Sociological Theory of Religion.* New York: Anchor, 1969.

Beyond Belief: Enlightenment 2.0. http://thesciencenetwork.org/BeyondBelief2/.

Bloch, Maurice. "Are Religious Beliefs Counter-Intuitive?" In *Radical Interpretation in Religion,* edited by Nancy K. Frankenberry, 129–46. Cambridge: Cambridge University Press, 2002.

Bloom, Paul. "Is God an Accident?" *Atlantic Monthly* 296, no. 5 (December 2005): 105–12.

Bloor, David. *Knowledge and Social Imagery.* 2nd ed. Chicago: University of Chicago Press, 1991.

Boghossian, Paul. *Fear of Knowledge: Against Relativism and Constructivism.* Oxford: Oxford University Press, 2006.

Bowler, Peter J., and Iwan Rhys Morus. *Making Modern Science: A Historical Survey.* Chicago: University of Chicago Press, 2005.

Boyer, Pascal. *The Naturalness of Religious Ideas: A Cognitive Theory of Religion.* Berkeley: University of California Press, 1994.

———. *Religion Explained: The Evolutionary Origins of Religious Thought.* New York: Basic Books, 2001.

Brooke, John Hedley. "Natural Theology." In *The History of Science and Religion in the Western Tradition*, edited by Gary B. Ferngren, 58–64. New York: Garland, 2000.

Brooke, John Hedley, and Geoffrey Cantor. *Reconstructing Nature: The Engagement of Science and Religion*. Edinburgh: T & T Clark, 1998.

Bulbulia, Joseph, Richard Sosis, Erica Harris, Russell Genet, Cheryl Genet, and Karen Wyman, eds. *The Evolution of Religion: Studies, Theories, and Critiques*. Santa Margarita, CA: Collins Foundation Press, 2008.

Burke, Kenneth. *The Rhetoric of Religion: Studies in Logology*. Berkeley: University of California Press, 1970.

Burkert, Walter. *Creation of the Sacred: Tracks of Biology in Early Religions*. Cambridge, MA: Harvard University Press, 1996.

Caporael, Linnda R. "The Evolution of Truly Social Cognition: The Core Configurations Model." *Personality and Social Psychology Review* 1, no. 4 (1997): 276–98.

Carroll, Joseph. *Literary Darwinism: Evolution, Human Nature, and Literature*. New York: Routledge, 2004.

Carroll, Robert P. *When Prophecy Failed: Cognitive Dissonance in the Prophetic Traditions of the Old Testament*. New York: Seabury Press, 1979.

Casanova, José. "Rethinking Secularization: A Global Comparative Perspective." *Hedgehog Review* 8, nos. 1–2 (2006): 7–22.

Cohn, Norman. *The Pursuit of the Millennium*. London: Secker & Warburg, 1957.

Collins, Francis S. *The Language of God: A Scientist Presents Evidence for Belief*. New York: Free Press, 2006.

Conway Morris, Simon. *Life's Solution: Inevitable Humans in a Lonely Universe*. Cambridge: Cambridge University Press, 2003.

Cook, Harold. *Matters of Exchange: Commerce, Medicine, and Science in the Dutch Golden Age*. New Haven: Yale University Press, 2007.

Cowie, Fiona. *What's Within? Nativism Reconsidered*. New York: Oxford University Press, 1999.

d'Aquili, Eugene G., and Andrew B. Newberg. "The Neuropsychological Basis of Religions, or Why God Won't Go Away." *Zygon* 33, no. 2 (1998): 187–201.

Darwin, Charles. *The Origin of Species by Means of Natural Selection: The Preservation of Favored Races in the Struggle for Life.* 1859. Edited by J. W. Burrow. London: Penguin, 1968.

Davie, Grace. *Religion in Britain since 1945: Believing without Belonging.* Oxford: Blackwell, 1994.

———. *Religion in Modern Europe: A Memory Mutates.* Oxford: Oxford University Press, 2000.

Davies, Paul. *The Mind of God: The Scientific Basis for a Rational World.* New York: Simon & Schuster, 1992.

Dawkins, Richard. *The God Delusion.* Boston: Houghton Mifflin, 2006.

———. *The Selfish Gene.* New York: Oxford University Press, 1976.

Deacon, Terrence W. *The Symbolic Species: The Co-evolution of Language and the Brain.* New York: Norton, 1997.

de Cruz, Helen, and Johan de Smedt. "The Role of Intuitive Ontologies in Scientific Understanding—The Case of Human Evolution." *Biology & Philosophy* 22 (2007): 351–68.

Dennett, Daniel C. *Breaking the Spell: Religion as a Natural Phenomenon.* New York: Viking, 2006.

———. *Consciousness Explained.* Boston: Little, Brown and Co., 1991.

———. *Darwin's Dangerous Idea: Evolution and the Meanings of Life.* New York: Simon & Schuster, 1995.

Descola, Phillipe, and Gísli Pálsson, eds. *Nature and Society: Anthropological Perspectives.* London: Routledge, 1996.

Diamond, Jared M. *Collapse: How Societies Choose to Fail or Succeed.* New York: Viking, 2005.

Dubuisson, Daniel. *The Western Construction of Religion: Myths, Knowledge, and Ideology.* Translated by William Sayers. Baltimore: Johns Hopkins University Press, 2003.

Durkheim, Émile. *The Elementary Forms of Religious Life.* 1912. Translated by Karen E. Fields. New York: Free Press, 1995.

Eliade, Mircea. *Patterns in Comparative Religion.* 1949. Translated by Rosemary Sheed. New York: Meridian Books, 1963.

Ferngren, Gary B., ed. *Science & Religion: A Historical Introduction.* Baltimore: Johns Hopkins University Press, 2002.

Festinger, Leon. *A Theory of Cognitive Dissonance.* Stanford: Stanford University Press, 1957.

Festinger, Leon, Henry W. Rieken and Stanley Schachter. *When Prophecy Fails.* Minneapolis: University of Minnesota Press, 1956.

Feyerabend, Paul. *Against Method: Outline of an Anarchistic Theory of Knowledge.* Atlantic Highlands, NJ: Humanities Press, 1975.

Fitzgerald, Timothy. *The Ideology of Religious Studies.* New York: Oxford University Press, 2000.

Flanagan, Owen. "Varieties of Naturalism." In *The Oxford Handbook of Religion and Science*, edited by Philip Clayton and Zachary Simpson, 430–52. Oxford: Oxford University Press, 2006.

Fleck, Ludwik. *Genesis and Development of a Scientific Fact.* 1935. Edited by Thaddeus J. Trenn and Robert K. Merton. Translated by Fred Bradley and Thaddeus J. Trenn. Chicago: University of Chicago Press, 1979.

Foucault, Michel. *The Archaeology of Knowledge.* Translated by A. M. Sheridan Smith. New York: Pantheon, 1972.

Frankenberry, Nancy K., ed. *Radical Interpretation in Religion.* Cambridge: Cambridge University Press, 2002.

Frazer, James. *The Golden Bough: A Study in Magic and Religion.* 2nd ed. London: Macmillan, 1900.

Geertz, Clifford. *The Interpretation of Cultures: Selected Essays.* New York: Basic Books, 1973.

Gigerenzer, Gerd. *Adaptive Thinking: Rationality in the Real World.* Oxford: Oxford University Press, 2000.

Gold, Daniel. *Aesthetics and Analysis in Writing on Religion: Modern Fascinations.* Berkeley: University of California Press, 2003.

Golinski, Jan. *Making Natural Knowledge: Constructivism and the History of Science.* Cambridge: Cambridge University Press, 1998.

Gombrich, E. H. *Art and Illusion: A Study in the Psychology of Pictorial Representation.* New York: Pantheon, 1960.

Gomperz, Theodor. *Greek Thinkers: A History of Ancient Philosophy.* Translated by L. Magnus and G. G. Berry. 4 vols. London: John Murray, 1901–12.

Goodman, Nelson. *Ways of Worldmaking.* Indianapolis: Hackett, 1978.

Goodwin, Brian C. *How the Leopard Changed Its Spots: The Evolution of Complexity.* New York: Charles Scribner's Sons, 1994.

Gopnik, Alison. "Explanation as Orgasm and the Drive for Causal Knowledge: The Function, Evolution, and Phenomenology of the Theory Formation System." In *Explanation and Cognition,* edited by Frank C. Keil and Robert A. Wilson, 299–324. Cambridge, MA: MIT Press, 2000.

Greenwood, John D. *The Disappearance of the Social in American Social Psychology.* Cambridge: Cambridge University Press, 2004.

Griffiths, Paul J. "Faith Seeking Explanation." Review of *Religion Explained,* by Pascal Boyer. *First Things,* no. 119 (2002): 53–57.

———. "*Tu Quoque.*" Review of *Manufacturing Religion,* by Russell T. McCutcheon. *First Things,* no. 81 (1998): 44–48.

———. "The Very Idea of Religion." *First Things,* no. 103 (2000): 30–35.

Guthrie, Stewart. *Faces in the Clouds: A New Theory of Religion.* Rev. ed. New York: Oxford University Press, 1995.

Hadot, Pierre. *The Veil of Isis: An Essay on the History of the Idea of Nature.* Translated by Michael Chase. Cambridge, MA: Harvard University Press, 2006.

Harmon-Jones, Eddie. "Toward an Understanding of the Motivation Underlying Dissonance Effects: Is the Production of Aversive Consequences Necessary?" In *Cognitive Dissonance: Progress on a Pivotal Theory in Social Psychology,* edited by Eddie Harmon-Jones and Judson Mills, 71–99. Washington, DC: American Psychological Association, 1999.

Harmon-Jones, Eddie, and Judson Mills, eds. *Cognitive Dissonance:*

Progress on a Pivotal Theory in Social Psychology. Washington, DC: American Psychological Association, 1999.

Harrison, Peter. "'Science' and 'Religion': Constructing the Boundaries." *Journal of Religion* 86, no. 1 (2006): 81–106.

Hart, David B. "Daniel Dennett Hunts the Snark." Review of *Breaking the Spell*, by Daniel Dennett. *First Things*, no. 169 (2007): 30–38.

Haught, John F. *Deeper than Darwin: The Prospect for Religion in the Age of Evolution.* Boulder, CO: Westview Press, 2003.

Hendriks-Jansen, Horst. *Catching Ourselves in the Act: Situated Activity, Interactive Emergence, Evolution, and Human Thought.* Cambridge, MA: MIT Press, 1996.

Hervieu-Léger, Danièle. "In Search of Certainties: The Paradoxes of Religiosity in Societies of High Modernity." *Hedgehog Review* 8, nos. 1–2 (2006): 59–68.

Hinde, Robert A. *Why Gods Persist: A Scientific Approach to Religion.* London: Routledge, 1999.

Hollinger, David. *Science, Jews, and Secular Culture: Studies in Mid-Twentieth-Century American Intellectual History.* Princeton: Princeton University Press, 1996.

Hume, David. *The Natural History of Religion.* 1757. In *Principal Writings on Religion.* Edited by J. C. A. Gaskin. Oxford: Oxford University Press, 1998.

Hutchins, Edwin. *Cognition in the Wild.* Cambridge, MA: MIT Press, 1995.

Janiak, Andrew. "Introduction." In *Philosophical Writings*, by Isaac Newton, ix–xxi. Cambridge: Cambridge University Press, 2004.

———. *Newton as Philosopher.* Cambridge: Cambridge University Press, 2008.

Johnson-Laird, Philip N., and Eldar Shafir. "The Interaction between Reasoning and Decision Making: An Introduction." *Cognition* 49, nos. 1–2 (1993): 1–9.

Kafka, Franz. "Leopards in the Temple." In *Parables and Paradoxes.*

Translated by Ernst Kaiser and Eithne Wilkins. New York: Schocken, 1961.

Kahneman, Daniel, Paul Slovic, and Amos Tversky, eds. *Judgment under Uncertainty: Heuristics and Biases.* Cambridge: Cambridge University Press, 1982.

Keane, Webb. "The Evidence of the Senses and the Materiality of Religion." *Journal of the Royal Anthropological Institute*, n.s., 14, no. S1 (2008): S110–S127.

Keil, Frank C., and Robert A. Wilson. "Explaining Explanation." In *Explanation and Cognition*, edited by Frank C. Keil and Robert A. Wilson, 1–18. Cambridge, MA: MIT Press, 2000.

Keller, Evelyn Fox. *The Century of the Gene.* Cambridge, MA: Harvard University Press, 2000.

Kessen, William. "The Transcendental Alarm." In *Historical Dimensions of Psychological Discourse*, edited by Kenneth J. Gergen and Carl F. Graumann, 263–74. Cambridge: Cambridge University Press, 1996.

Kirschner, Suzanne R. "Sources of Redemption in Psychoanalytic Developmental Psychology." In *Historical Dimensions of Psychological Discourse*, edited by Kenneth J. Gergen and Carl F. Graumann, 193–203. Cambridge: Cambridge University Press, 1996.

Knight, Christopher C. *Wrestling with the Divine: Religion, Science, and Revelation.* Minneapolis: Fortress Press, 2001.

Knorr-Cetina, Karin. *Epistemic Cultures: How the Sciences Make Knowledge.* Cambridge, MA: Harvard University Press, 1999.

———. *The Manufacture of Knowledge: An Essay on the Constructivist and Contextual Nature of Science.* Oxford: Pergamon Press, 1981.

Kuhn, Thomas S. *The Structure of Scientific Revolutions.* Chicago: University of Chicago Press, 1962.

Lakoff, George, and Mark Johnson. *Metaphors We Live By.* Chicago: University of Chicago Press, 1980.

Latour, Bruno. *The Pasteurization of France.* Translated by Alan

Sheridan and John Law. Cambridge, MA: Harvard University Press, 1988.

———. *Science in Action: How to Follow Scientists and Engineers through Society*. Cambridge, MA: Harvard University Press, 1987.

———. "'Thou Shall Not Freeze-Frame,' or, How Not to Misunderstand the Science and Religion Debate." In *Science, Religion, and the Human Experience*, edited by James D. Proctor, 27–48. Oxford: Oxford University Press, 2005.

Latour, Bruno, and Peter Weibel, eds. *Iconoclash: Beyond the Image Wars in Science, Religion, and Art*. Translated by Charlotte Bigg et al. Cambridge, MA: MIT Press, 2002.

Lawson, E. Thomas, and Robert N. McCauley. *Rethinking Religion: Connecting Cognition and Culture*. Cambridge: Cambridge University Press, 1990.

Lewontin, Richard. *The Triple Helix: Gene, Organism, and Environment*. Cambridge, MA: Harvard University Press, 2000.

Lieberman, Philip. *Toward an Evolutionary Biology of Language*. Cambridge, MA: Harvard University Press, 2006.

Lightman, Alan, and Owen Gingerich. "When Do Anomalies Begin?" *Science*, n.s., 255, no. 5045 (February 7 1992): 690–95.

Lipton, Peter. "Science and Religion: The Immersion Solution." In *Realism and Religion: Philosophical and Theological Perspectives*, edited by Andrew Moore and Michael Scott, 52–79. Aldershot: Ashgate, 2007.

Lurie, Alison. *Imaginary Friends*. New York: Coward-McCann, 1967.

Mahmood, Saba. *The Politics of Piety: The Islamic Revival and the Feminist Subject*. Princeton: Princeton University Press, 2005.

Masuzawa, Tomoko. "In Nature's Trail." Review of *Creation of the Sacred*, by Walter Burkert. *Method & Theory in the Study of Religion* 10, no. 1 (1998): 106–14.

———. *The Invention of World Religions, or, How European Universalism Was Preserved in the Language of Pluralism*. Chicago: University of Chicago Press, 2005.

McCauley, Robert N. "The Naturalness of Religion and the Unnaturalness of Science." In *Explanation and Cognition*, edited by Frank C. Keil and Robert A. Wilson, 61–85. Cambridge, MA: MIT Press, 2000.

McCutcheon, Russell T. *The Discipline of Religion: Structure, Meaning, Rhetoric.* London: Routledge, 2003.

———. *Manufacturing Religion: The Discourse on Sui Generis Religion and the Politics of Nostalgia.* New York: Oxford University Press, 1997.

Melser, Derek. *The Act of Thinking.* Cambridge, MA: MIT Press, 2004.

Melton, J. Gordon. "Spiritualization and Reaffirmation: What Really Happens When Prophecy Fails." In *Expecting Armageddon: Essential Readings in Failed Prophecy*, edited by Jon R. Stone, 145–57. New York: Routledge, 2000.

Meyer, Leonard B. *Emotion and Meaning in Music.* Chicago: University of Chicago Press, 1956.

Miller, Kenneth. *Finding Darwin's God: A Scientist's Search for Common Ground between God and Evolution.* New York: Cliff Street Books, 1999.

Mynatt, Clifford R., Michael E. Doherty, and Ryan D. Tweney. "Confirmation Bias in a Simulated Research Environment: An Experimental Study of Scientific Inference." *Quarterly Journal of Experimental Psychology* 29, no. 1 (1977): 85–95.

Nagel, Thomas. "The Fear of Religion." Review of *The God Delusion*, by Richard Dawkins. *The New Republic* 235, no. 17 (23 October 2006): 25–29.

Newton, Isaac. *Philosophical Writings*, edited by Andrew Janiak. Cambridge: Cambridge University Press, 2004.

Niedenthal, Paula M. "Embodying Emotion." *Science*, n.s., 316, no. 5827 (18 May 2007): 1002–5.

Nielsen, Kai. *Naturalism and Religion.* Amherst, NY: Prometheus Books, 2001.

Nietzsche, Friedrich. *On the Genealogy of Morals.* 1887. Translated by Walter Kaufmann and R. J. Hollingdale. New York: Vintage, 1989.

Nisbett, Richard E., and Lee Ross. *Human Inference: Strategies and Shortcomings of Social Judgment.* Englewood Cliffs, NJ: Prentice-Hall, 1980.

Noble, David F. *World without Women: The Christian Clerical Culture of Western Science.* New York: Knopf, 1992.

Noë, Alva. *Action in Perception.* Cambridge, MA: MIT Press, 2004.

Núñez, Rafael, and Walter J. Freeman, eds. *Reclaiming Cognition: The Primacy of Action, Intention, and Emotion.* Thorverton: Imprint Academic, 1999.

Olafson, Frederick. *Naturalism and the Human Condition: Against Scientism.* London: Routledge, 2001.

Olson, Richard. *Science and Religion, 1450–1900: From Copernicus to Darwin.* Baltimore: Johns Hopkins University Press, 2004.

Orsi, Robert A. *Between Heaven and Earth: The Religious Worlds People Make and the Scholars Who Study Them.* Princeton: Princeton University Press, 2005.

Oyama, Susan. *The Ontogeny of Information: Developmental Systems and Evolution.* 2nd ed. Durham, NC: Duke University Press, 2000.

Peacocke, A. R. *Theology for a Scientific Age: Being and Becoming—Natural, Divine, and Human.* Minneapolis: Fortress Press, 1993.

Peterson, Gregory R. "Demarcation and the Scientistic Fallacy." *Zygon* 38, no. 4 (2003): 751–61.

Pfeifer, Rolf, and Josh C. Bongard. *How the Body Shapes the Way We Think: A New View of Intelligence.* Cambridge, MA: MIT Press, 2006.

Pickering, Andrew. *The Mangle of Practice: Time, Agency, and Science.* Chicago: University of Chicago Press, 1995.

Pinker, Steven. *The Blank Slate: The Modern Denial of Human Nature.* New York: Viking, 2002.

———. *How the Mind Works.* New York: Norton, 1997.

Polkinghorne, John. *Exploring Reality: The Intertwining of Science and Religion.* New Haven: Yale University Press, 2005.

Port, Robert F., and Timothy van Gelder, eds. *Mind as Motion: Ex-*

plorations in the Dynamics of Cognition. Cambridge, MA: MIT Press, 1995.

Proctor, Robert N. *Value-Free Science? Purity and Power in Modern Knowledge.* Cambridge, MA: Harvard University Press, 1991.

Pyysiäinen, Ilkka. *How Religion Works: Towards a New Cognitive Science of Religion.* Leiden: Brill, 2001.

Richardson, J. T. "Experiencing Research on New Religions and 'Cults': Practical and Ethical Considerations." In *Experiencing Fieldwork: An Inside View of Qualitative Research,* edited by William Shaffir and Robert A. Stebbins, 62–71. Newbury Park, CA: Sage, 1991.

Richardson, Robert C. *Evolutionary Psychology as Maladapted Psychology.* Cambridge, MA: MIT Press, 2007.

Richerson, Peter J., and Robert Boyd. *Not by Genes Alone: How Culture Transformed Human Evolution.* Chicago: University of Chicago Press, 2005.

Rockwell, W. Teed. *Neither Brain nor Ghost: A Nondualist Alternative to the Mind-Brain Identity Theory.* Cambridge, MA: MIT Press, 2005.

Rorty, Richard. *Philosophy and the Mirror of Nature.* Princeton, NJ: Princeton University Press, 1979.

Saler, Benson. *Conceptualising Religion: Immanent Anthropologists, Transcendent Natives, and Unbounded Categories.* Leiden, The Netherlands: Brill, 1993.

Schüler, Sebastian. "Embodied Cognition and Ritual Synchronization—Adaptive Dynamics of Interaction in Glossolalia." Unpublished paper originally presented at the Conference on Symbolization in Religion, International Association for the Cognitive Science of Religion, June 2007.

Segal, Robert A. "All Generalizations Are Bad: Postmodernism on Theories." *Journal of the American Academy of Religion* 74, no. 1 (2006): 157–71.

———. *Explaining and Interpreting Religion: Essays on the Issue.* New York: Peter Lang, 1992.

Shapin, Steven. *The Scientific Life: A Moral History of a Late Modern Vocation.* Chicago: University of Chicago Press, 2008.

Shapin, Steven, and Simon Schaffer. *Leviathan and the Air Pump: Hobbes, Boyle, and the Experimental Life.* Princeton: Princeton University Press, 1985.

Slingerland, Edward G. *What Science Offers the Humanities: Integrating Body and Culture.* Cambridge: Cambridge University Press, 2008.

Slone, D. Jason. *Theological Incorrectness: Why Religious People Believe What They Shouldn't.* Oxford: Oxford University Press, 2004.

Smart, Ninian. *The Science of Religion and the Sociology of Knowledge: Some Methodological Questions.* Princeton: Princeton University Press, 1973.

Smith, Barbara Herrnstein. *Belief and Resistance: Dynamics of Contemporary Intellectual Controversy.* Cambridge, MA: Harvard University Press, 1997.

———. "The Complex Agony of Injustice." *Cardozo Law Review* 13, no. 4 (1991): 1273–75.

———. *Contingencies of Value: Alternative Perspectives for Critical Theory.* Cambridge, MA: Harvard University Press, 1988.

———. *On the Margins of Discourse: The Relation of Literature to Language.* Chicago: University of Chicago Press, 1978.

———. "Reply to an Analytic Philosopher." *South Atlantic Quarterly* 101, no. 1 (2002): 229–42.

———. *Scandalous Knowledge: Science, Truth and the Human.* Edinburgh: University of Edinburgh Press, 2005; Durham, NC: Duke University Press, 2006.

Smith, Huston. *Why Religion Matters: The Fate of the Human Spirit in an Age of Disbelief.* New York: HarperCollins, 2001.

Smith, Jonathan Z. "Religion, Religions, Religious." In *Critical Terms for Religious Study*, edited by Mark C. Taylor, 269–84. Chicago: University of Chicago Press, 1998.

Smith, Roger. *Being Human: Historical Knowledge and the Creation of Human Nature.* Manchester, UK: Manchester University Press, 2007.

Smolin, Lee. *The Trouble with Physics: The Rise of String Theory, the Fall of Science, and What Comes Next.* Boston: Houghton Mifflin Co., 2006.

Spilka, Bernard, Ralph W. Hood Jr., Bruce Hunsberger, and Richard Gorsuch. *The Psychology of Religion: An Empirical Approach.* 3rd ed. New York: Guilford Press, 2003.

Stenmark, Mikael. *Scientism: Science, Ethics, and Religion.* Aldershot: Ashgate, 2001.

Sterelny, Kim. *Thought in a Hostile World: The Evolution of Human Cognition.* Malden, MA: Blackwell, 2005.

Thelen, Esther, and Linda B. Smith. *A Dynamic Systems Approach to the Development of Cognition and Action.* Cambridge, MA: MIT Press, 1994.

Thomson, Keith. *Before Darwin: Reconciling God and Nature.* New Haven: Yale University Press, 2005.

Tooby, John, and Leda Cosmides. "The Psychological Foundations of Culture." In *The Adapted Mind: Evolutionary Psychology and the Generation of Culture,* edited by Jerome H. Barkow, Leda Cosmides, and John Tooby, 19–136. New York: Oxford University Press, 1992.

Tylor, Edward B. *Religion in Primitive Culture* [originally *Primitive Culture*]. 1871. New York: Harper, 1958.

van Gelder, Timothy, and Robert F. Port. "It's About Time: An Overview of the Dynamical Approach to Cognition." In *Mind as Motion: Explorations in the Dynamics of Cognition,* edited by Robert F. Port and Timothy van Gelder, 1–43. Cambridge, MA: MIT Press, 1995.

Varela, Francisco J., Evan Thompson, and Eleanor Rosch. *The Embodied Mind: Cognitive Science and Human Experience.* Cambridge, MA: MIT Press, 1991.

Veyne, Paul. *Did the Greeks Believe in Their Myths? An Essay on the Constitutive Imagination.* Chicago: University of Chicago Press, 1988.

Vilenkin, Alex. *Many Worlds in One: The Search for Other Universes.* New York: Hill and Wang, 2006.

Wallace, Anthony F. C. *Religion: An Anthropological View.* New York: Random House, 1966.

Weber, Bruce H., and David J. Depew. "Developmental Systems, Darwinian Evolution, and the Unity of Science"; in Susan Oyama, Paul E. Griffiths, and Russell D. Gray, eds., *Cycles of Contingency: Developmental Systems and Evolution.* Cambridge, MA: MIT Press, 2001.

Wexler, Bruce E. *Brain and Culture: Neurobiology, Ideology, and Social Change.* Cambridge, MA: MIT Press, 2006.

Wheeler, Michael. *Reconstructing the Cognitive World: The Next Step.* Cambridge, MA: MIT Press, 2005.

Wiebe, Donald. *The Politics of Religious Studies: The Continuing Conflict with Theology in the Academy.* New York: St. Martin's Press, 1998.

Wilber, Ken. *The Marriage of Sense and Soul: Integrating Science and Religion.* New York: Random House, 1998.

Williams, Raymond. "Nature." In *Keywords: A Vocabulary of Culture and Society.* Rev. ed. London: Fontana, 1983.

Wilson, David Sloan. *Darwin's Cathedral: Evolution, Religion, and the Nature of Society.* Chicago: University of Chicago Press, 2002.

Wilson, Edward O. *Consilience: The Unity of Knowledge.* New York: Knopf, 1998.

Wolpert, Lewis. *The Unnatural Nature of Science.* Cambridge, MA: Harvard University Press, 1993.

Woodward, James, and Fiona Cowie. "The Mind Is Not (Just) a System of Modules Shaped (Just) by Natural Selection." In *Contemporary Debates in Philosophy of Science,* edited by Christopher Hitchcock, 312–34. New York: Blackwell, 2004.

Zygmunt, Joseph F. "When Prophecies Fail: A Theoretical Perspective on the Comparative Evidence." *American Behavioral Scientist* 16, no. 2 (1972): 245–68.

Index

adaptiveness, of religion, 34–35, 108, 112, 160n28, 169n16
agents: cognitive device for detecting, 38–39, 67–68, 160nn24–25; supernatural, 57, 67–68, 89–94, 160n25, 164n15, 167n40
ancestors, beliefs about, 38, 39, 91–92, 164n15
animism, 39, 85, 134
anthropology, and religious studies, xii, 20, 30, 31, 54, 55–57, 62, 84, 166n34
anthropomorphism, 39–40, 42, 134
Aquinas, Thomas, 96, 171n6
Asad, Talal, 157n8
atheism, atheists, 22–23, 71, 84, 88, 115, 122, 123, 128, 148–149, 153n15, 170n24
Atran, Scott, *In Gods We Trust*, xiii, 59–74, 80–90, 160n26, 161n35, 163n2, 163–164nn10–12, 167n41, 169n16; on alternative accounts of religion, 66–69; on beliefs, 70–72; on benefits of religion, 80–88; on cognitive universals, 63–64; commitments to evolutionary psychology, 61–65; on Itza' Mayan ecology, 84–85; on neurotheology, 72–74; rationalist-realist bias in, 70–74
Attachment Theory, 67–68, 163n10
Augustine, 96; *Confessions*, 66
Barrett, Justin L., 160n24, 170n24
Beckford, James, 20–21
belief, beliefs: alternative conceptions of, xi–xii, 14–19, 69–80, 165n23; cross-cultural recurrence of, 63–64; Daniel Dennett on, 76–80, 165nn22–24; incoherence of, 18–19, 75–76, 127–128, 154–155nn24–25; persistence of, 3–10, 15–16, 57, 134, 152nn8–9, 153n14
Bell, Catherine M., 75, 162n46, 170n25
Berger, Peter, 155n26
biology, 36, 37, 62, 141, 169n20, 176n34; and accounts of religion, xiii, 49, 50, 59–60, 62, 86, 177n34 (*see also* New Naturalism); theological readings of, 100–101, 103, 113. *See also* evolutionary theory
Bloch, Maurice, 91–92, 94, 167n44
Bloor, David, 154n17, 170n29, 177n35
Boghossian, Paul, 170n29
Bowler, Peter J., 159n22, 174n19
Boyd, Robert, 169n20
Boyer, Pascal, *Religion Explained*, xiii, 33–39, 41–48, 50–51, 54–57, 107, 112, 137; on death and ideas of immortality, 34, 43–45, 81–82; on morality and gods, 34, 45–48; on other, prior accounts of religion, 33–38, 41, 56, 155n25, 161n30, 161n35, 169n16;

Boyer, Pascal (continued)
religion-science contrasts in, 120, 128–130, 131–132, 133, 138, 172nn10–11, 173n15; on rituals, 54–57, 109, 163n47; on scientific explanation, 36, 37–38, 56–57, 159n21, 159n23, 160n25, 162n39; on supernatural agents, 38–39, 42, 89–90; theological critiques of, 112, 114, 118–120
Brooke, John Hedley, 17
Buddhism, 124, 133
Burke, Kenneth, 168n6
Burkert, Walter, *Creation of the Sacred:* as naturalist account of religion, xiii, 49–54, 68, 109, 110, 112, 165n22; reception of, 54, 162n39, 162n43, 175n25, 176n34

Caporael, Linnda R., 154n21
Carroll, Joseph, 176n32
Carroll, Robert, 21–22
Casanova, José, 166n28
Catholic Church, 129, 132, 156n7, 172n10
Chomsky, Noam, 158n14, 163n9
Christianity: and concept of religion, 30, 156–157nn6–7; and heterogeneous beliefs, 18, 76; vis-à-vis science, 95, 97, 101, 103, 110, 132, 173n14 (*see also* New Natural Theology)
classicists, and religious studies, 49, 56–57, 133
cognition, human: alternative views of, xi–xii, 10–14, 33, 36–37, 39, 49, 72–75, 76, 153n16, 158–159nn19–20; general tendencies of, xiv, 1, 5–10, 23–24, 39–43, 85, 119–120, 132–136, 148, 160n28, 173–174n16; social aspects of, 15–16, 45–46, 154nn20–21. *See also* belief; cognitive science
cognitive conservatism, 5–10, 16–18, 21, 134, 148, 174n17
cognitive consonance, 14, 16, 102–103
cognitive dissonance, 4, 5, 6–7, 15, 21–22, 93, 105, 107, 152–153n9, 153n14
cognitive science, 8, 12, 31, 36, 50, 58, 76 (*see also* neuroscience; psychology); and explanations of religion, xii–xiii, 12, 19, 25–26, 32–48, 54–57, 60–96, 119–120, 157n11 (*see also* neurotheology; New Naturalism)
Cohn, Norman, 151n2
Collins, Francis, 167n4
constructivism. *See* epistemology, constructivist-pragmatist views of
Cosmides, Leda, 93, 107, 157n13, 164n12, 176n31
cosmologies, personal, 127, 128, 148, 149
cosmology, 13, 98, 99, 123–124, 171n3
counterintuitiveness, 70–71, 91–93, 131, 136–137
Cowie, Fiona, 158n20, 160n23, 163n9, 164n13
cults, 16, 19
culture: concept of, 9, 27; religion and, 9, 50, 94

d'Aquili, Eugene G., 72, 164n17
Darwin, Charles, 41, 110
Darwinism: and "Darwinian thinking," 110–111; and explanations of religion, 96, 106–120 (*see also* New Naturalism); theistic and theological responses to, 17, 95–96, 101–102, 106–120. *See also* evolutionary theory
Davies, Paul, 167n4
Dawkins, Richard, 110, 165n24, 169n16, 169n19, 169n24, 171n4
Deacon, Terrence W., 163n9
death, and religious belief, 43–45, 81–82
Dennett, Daniel C., *Breaking the Spell*, 160n26, 170n28; on "anti-Darwinians," 140–145, 162n43; on belief, 76–80, 115, 164n15, 165nn22–24, 166n38; and critics of religion, 169–170n24; and programmatic naturalism, 139, 146–147, 157n9, 158n15, 164n14, 172n9; on science studies, 175n25, 177n35
Depew, David J., 169n20
Descola, Phillipe, 171n6
Diamond, Jared, 84
Douglas, Mary, 170n24
Dubuisson, Daniel, 156n2
Durkheim, Émile, 16, 33, 162n46

ecology, and supernatural beliefs, 84–85
economics, 59–60. *See also* rational-choice theory

Eliade, Mircea, 28, 31, 33, 51, 155n32
embodiment: of cognitive processes, xii, 11–12, 153n16; of religious belief, 78, 161n38, 166n25
Enlightenment (movement): and New Naturalism, 70, 114, 126, 164n14, 170n24; and science-religion connections, 132
epistemology, 10, 26, 70, 105, 114, 124, 141, 152n8, 152n9, 164n13; constructivist-pragmatist views of, xii, 13–14, 72–75, 116–117, 154n17, 170n29; dualisms in, 168n13; rationalist-realist views of, 72–75
evolutionary psychology, 50, 93, 107, 109, 145, 158n18, 159n21, 159n23, 160n28, 176n32; and explanations of religion, xiii, 26, 31, 35, 50, 61 (*see also* New Naturalism); mental mechanisms and models in, 37–38, 42, 58, 62, 63, 86, 138
evolutionary theory: and religion, 17, 59–60, 62, 87, 98–103, 106–120, 123, 145 (*see also* New Naturalism); and Intelligent Design theory, 123. *See also* Darwinism; group selection; Natural Selection
exceptionalism, scientific, xv, 131, 133, 146
experience: mystical, 72, 74, 127; religious, 22–28, 55–56, 90, 94, 105, 106, 109, 128
explanation, views of, 33, 37–38, 41–42, 50, 54, 55–58, 67–68, 123, 138, 159n21

Festinger, Leon, and cognitive dissonance theory, 21, 152nn8–9. See also *When Prophecy Fails*
Feyerabend, Paul, 154n17
Fitzgerald, Timothy, 156n5, 156n7
Flanagan, Owen, 171n1
Fleck, Ludwik, 13–14, 16, 20–21, 154n17, 174n17
Foucault, Michel, 53, 154n17, 177n35
Frazer, James G., *The Golden Bough*, 41, 161n30
Freeman, Walter J., 153n16, 164n13
Freud, Sigmund, 20, 52, 113
fundamentalism, religious, 108, 136

game theory, xiii, 47, 49, 62
Geertz, Clifford, 31, 175n30
genes, 109–111, 169n20
Genesis, 98
Gigerenzer, Gerd, 153n13
Gingerich, Owen, 153n10
God, 98, 103, 110, 163n10, 166n38
gods: explanations of, 35, 38–40, 42, 46–47, 51–52, 64, 67–68, 70, 72, 90, 163–164n11; and immortality, 81–82; and morality, 83–84, 87–88; persistence of, 21, 86, 94, 107, 109. *See also* agents, supernatural
Gold, Daniel, 156n4
Golinski, Jan, 154n17
Gomperz, Theodor, 41, 49, 160n29
Goodman, Nelson, 154n17
Goodwin, Brian C., 169n20
Gopnik, Alison, 138–139, 174n22
Götterdämmerung (Richard Wagner), 110
Greenwood, John D., 154n21
Griffiths, Paul J., 29–30, 33, 118–120, 122–123, 155n32, 156nn6–7, 170n30
group selection, 161n35, 169n16

Hadot, Pierre, 171n5
Hamlet (character), 128
Harrison, Peter, 173n13
Hart, David B., 168n11
Haught, John, *Deeper than Darwin:* on Darwinian explanations of religion, 106–117, 118, 162n43, 169n24, 173n14; as reconciliation of science and religious belief, 95–104
Hendriks-Jansen, Horst, 153n16, 158–159nn19–20
hermeneutics, theological, 98, 103
history, in accounts of religion, 42–43, 53, 56–57, 85, 88, 132, 134–136
Hollinger, David, 173n14
human nature, 50, 130, 136, 140, 176n32
humanities, xv, 32, 49, 124, 139; vis-à-vis sciences, 140–146, 176nn33–34
Hume, David, *The Natural History of Religion*, 27, 33, 40, 49, 161n32
Hutchins, Edwin, 153n16, 159n20

immortality, beliefs about, 35, 43–45, 81–83, 97, 161n34, 164n15
Intelligent Design theory, 96, 99, 123

International Association for the Cognitive Science of Religion, 157n11
intuitiveness, 8, 43, 45, 64, 70–71, 91–93, 164n15, 174n16. *See also* counter-intuitiveness

Janiak, Andrew, 173n13
Johnson, Mark, 64

Kafka, Franz, 17
Keane, Webb, 165n25, 175n27
Keech, Marion, 2–4, 18, 75, 151n5. *See also* Searchers
Keller, Evelyn Fox, 169n20
Knight, Christopher C., 167n1
Knorr-Cetina, Karin, 154n17, 156n33
knowledge: alternative views of, xi–xii, 10–13, 64–65, 154n17; normative claims about, 124–125, 132, 143, 148. *See also* epistemology; scientific knowledge
Kuhn, Thomas S., 7, 174n17, 177n35

Lakoff, George, 64
language: acquisition of, 64–65, 163n9; and belief-persistence, 154n20; Chomskian theory of, 107; religious and theological, 97–98, 102
Latour, Bruno, 154n17, 156n33, 168n10, 177n35
Lawson, E. Thomas, 30–31, 107, 157n14, 163n47, 163n9
Lewontin, Richard, 169n20
Lieberman, Philip, 163n9
Lipton, Peter, 168n10
literary studies, 124, 144, 145, 176n32, 176n34
Lucretius, *De Rerum Natura*, 27, 112
Lurie, Alison, *Imaginary Friends*, 151n7

Mahmood, Saba, 165n25
Malagasy, 92
Marx, Karl, 113, 159n21
Masuzawa, Tomoko, 157n8, 162n43
McCauley, Robert N., 30–31, 107, 130–138, 147, 157n14, 163n47, 163n9, 166n29, 172nn11–12, 173n16
McCutcheon, Russell T., 155n32, 156n5, 170n30
Melser, Derek, 153n16
memes, 160n25, 165n24
millenarian movements, 1–4, 15–16, 151n2, 152n9. *See also* Searchers
Miller, Kenneth, 167n1
modularism, in cognitive science, xii, 35, 36, 42, 62, 67, 75, 82, 86–87, 90, 158nn19–20
morality, and religious belief, 34, 45–48, 80–81, 83–84, 87–88. *See also* ethics, matters of
Morris, Simon Conway, 167n1
mortality, 34, 43–45, 81–83
Morus, Iwan Rhys, 159n22, 174n19
mysticism. *See* experience, mystical
myths, 50, 75, 85, 107, 112, 165n20

Nagel, Thomas, 171n4
natural, concept of, 125–126
Natural Selection, as quasi-agent, 91, 110, 160n28
natural theology, 96, 167n4, 168n6. *See also* New Natural Theology
naturalism: and human phenomena, 28, 48–50, 53–54, 140–145, 146; "metaphysical" vs. "methodological," 121–125, 171n4; and religious studies, 22–23, 26–33, 39–43, 48–50, 122–123, 140–142; science and, 121–125, 147–148. *See also* naturalists; New Naturalism
naturalists, 100, 128, 146; ethical obligations of, 113–114; and reflexivity, 115–120, 147, 149
nature, as concept and construct, 125–128
neuroscience, xii, 10, 36, 37, 152n9, 164n13
neurotheology, 72–74
New Natural Theology, xiii–xv, 22, 95–105, 134
New Naturalism: as approach in religious studies, xii–xiii, xiv, 21, 31, 32, 57–58, 157n11, 161nn34–35; assessments of religion in, 80–88; confining commitments of, 61–69, 86–88, 93, 169n16; and earlier naturalist accounts of religion, 33–37, 41–43, 113–114; explanatory ambitions of, 43–58; as preemptive naturalism, 31–32, 139–146; and rival contemporary

accounts, 66–69, 91–94, 161–162n38; science-religion contrasts in, 77–80, 120, 128–139; theological responses to, 106–120; views of belief in, 59, 61, 69–80, 160n25
Newberg, Andrew B., 72, 164n17
Newton, Isaac, 132, 173n13
Niedenthal, Paula M., 153n16
Nielsen, Kai, 171n1
Nietzsche, Friedrich, 53; *On the Genealogy of Morals*, 25, 40, 49, 147, 161n32
Noble, David F., 173nn13–14
Noë, Alva, 153n16
Núñez, Rafael, 153n16, 164n13

Olson, Richard, 173n13
ontology, 104, 122–125
Orsi, Robert A., 76, 165n25
Oyama, Susan, 153n16, 169n20

Paley, William, 96
pantheism, 104, 124
"parasitism," cognitive, 47, 137–138
Peacocke, Arthur, 95, 97
philosophy, philosophers: 36, 49, 121–123, 125, 133, 142, 143, 145; and New Naturalism, 77–79, 117; of science, 10, 11, 13, 26, 37, 134, 147, 177n35
physics, 99, 123, 135, 171n3
Pickering, Andrew, 154n17, 177n35
Pinker, Steven, 93, 158n15, 164n12, 176n31
Polkinghorne, John, 95, 96, 99
polytheism, 104
Port, Robert F., 153n16, 158n19
"postmodernism," invocation of, 115, 126, 146
pragmatist epistemology, and authority of science, 116–117. *See also* epistemology, constructivist-pragmatist views of
Presbyterianism, 147–148, 177n37
projection (psychological), 39, 68, 84, 85
prophecy: biblical, 21–22; millenarian, 2–4, 16; vis-à-vis scientific prediction, 3–4, 5–7
psychology, psychologists, 28, 41, 62, 70, 73, 144; developmental, 11, 36, 138, 173n14; of religion, 175n29; social, 1–7, 15–16, 47, 152n9, 154n21. *See also* evolutionary psychology
Pyysiäinen, Ilkka, 157n9, 158n15, 164n15, 168n14

rational-choice theory, 45, 62, 85, 152n9
rationalism, xii, 19, 61, 69–80, 164n13
realism, reality, 72–75, 78–79, 124
reconciliations of science and religion, 95–96, 100, 104–105, 108, 167n4, 168n6. *See also* New Natural Theology
reflexivity, 115–116, 117–120, 149
relativism, 69, 73, 114, 164n12, 170n29
religion: ancient, 49–54, 112, 160–161n29; biological assessments of, 59–61, 80–88; as concept, 27–28, 30, 89–94, 105, 128–129, 132, 156nn6–7, 166n40; conflicts with science, 104–105; connections and continuities with science, 23–24, 132–133, 147–149, 172n10, 172–173nn12–14; contrasts of with science, 77–80, 120, 128–139; and "dread of mortality," 34, 43–45, 81–83; fundamentalist, 108, 136; and morality, 34, 45–48, 80–81, 87–88; naturalistic accounts of, 26–33, 48–54, 57–58, 59–74, 80–96, 106–120 (*see also* New Naturalism); persistence of, 19–20, 80–88, 94; reconciliations with science, 95–98, 100, 104–105, 108, 167n4, 168n6 (*see also* New Natural Theology); survival vis-à-vis science, 7, 60, 80–81, 136–137, 147, 174n21 (*see also* secularization); value of, 114–115. *See also* belief; Christianity; experience, religious; gods; prophecy; ritual; supernatural concepts; theism; theology; worship
religious studies, xv, 27–31, 56–57, 106, 139–142, 155–156n32, 156n5, 157–158n14, 170n24
Richardson, Robert C., 160n23, 163n2
Richerson, Peter J., 169n20
ritual, 35, 50, 54–57, 59, 84, 107, 108, 109, 158n14, 162n45, 163n47, 163n9, 169n16

Rockwell, W. Teed, 153n16
Rorty, Richard, 154n17
Rosch, Eleanor, 153n16
Rosicrucianism, 168n6

Saler, Benson, 156n2, 157n8
Schüler, Sebastian, 161n38
science: concept of, xi–xii, 105, 121–122, 128–129, 132, 135; conflicts with religion, 104–105, 122–123; connections with religion, xii, 23–24, 80, 117, 120, 132–133, 147–148, 172n10, 172–173nn12–14; connections with technology, 135, 174n19; contrasts to religion, xiv–xv, 7, 60, 77–80, 80–81, 120, 128–139; exceptionalism, xv, 131, 133, 146; explanation in, 33–34, 37–38, 123; general cognitive tendencies in, 6–7, 119–120, 132–136, 153n10, 160n28, 174n17; methodological naturalism and, 121–125, 147–148; naturalistic accounts of, 117, 138–139, 146–147; as "parasitic," 137–138; as "unnatural," 120, 129–139, 147; value or appeal vis-à-vis religion, 60, 80–81, 136–137, 147–148, 174n21. *See also* scientific knowledge; social sciences
science studies, xii, 23, 117, 134, 147–148, 170n29, 177n35
scientific knowledge, xi, 19–20, 23–24, 116–117, 139, 146–147, 177n35; authority of, 116–117, 128, 156n33; self-critique of, 147; vis-à-vis religious belief, xii, 24, 80, 95–96, 96–105, 118–119, 147–148, 167n4, 174n21. *See also* science; epistemology
scientism, 31–32, 125, 157n12, 158n14
Searchers (millenarian group), 2–4, 15–16, 18, 153n11
secularization, 19–21, 60, 80–81, 174n21
Segal, Robert A., 156n4, 170n25, 172n7
self-refutation, charge of, xiv, 23, 109, 115–116, 117–118, 118–120, 170–171nn29–30
shamanism, 147–148, 177n37
Shapin, Steven, 174n19
Slingerland, Edward G., 146, 157n13, 177n35
Slone, D. Jason, 161n29, 172n11
Smart, Ninian, 175n26
Smith, Huston, 157n12
Smith, Roger, 161n30
Smolin, Lee, 171n3
social sciences, 37, 49, 139–146
sociobiology, 53, 145, 161n30, 177n34
spirits, beliefs about, 38, 39, 75, 84–85, 92, 164n15
Sterelny, Kim, 159n20
supernatural, as concept and term, 89–91, 94, 126–127, 160n29, 171–172n6
supernatural concepts (beings, forces, realms), New Naturalist accounts of, 20, 38, 42, 47, 67–68, 70, 71, 80, 82, 93, 160n25, 164n15, 165n23, 166n38
symbols, and symbolic behavior, 51, 54, 64

theism, theists, 6, 22, 96, 123, 149, 153
theology, theologians, xv, 6, 17, 22–23, 29–30, 95–97, 115, 118, 121, 122, 124, 139, 168n11; critiques of naturalist explanations of religion by, 106–120; responses to science by, xiii–xiv, 6, 17, 22, 95–105. *See also* natural theology; New Natural Theology
Thompson, Evan, 153n16
Thomson, Keith, 167n2
Tooby, John, 93, 107, 157n13, 164n12, 176n31
tu quoque (you, too), charge of, 22–24, 118, 123, 149, 155–156n32
Tübingen School, 170n29
Two Cultures, ideology of, xv, 139–146, 176n33
Tylor, Edward B., *Primitive Culture*, 41

universalist attributions, 63–65, 70–71, 88, 92–93

van Gelder, Timothy, 153n16, 158n19
Varela, Francisco J., 153n16
Veyne, Paul, 75, 89, 165n20
Vilenkin, Alex, 171n3

Wallace, Anthony, 20
Weber, Bruce H., 169n20
Weber, Max, 16, 33, 49
Wexler, Bruce E., 152n9
Wheeler, Michael, 153n16

When Prophecy Fails (Festinger et al.), 3–5, 152n8
Williams, Raymond, 171n5
Wilson, David Sloan, 161n35, 169n16
Wilson, Edward O., 32, 146, 157n13, 164n14, 176n31, 177n35
Wolpert, Lewis, 130, 131–132, 135, 137, 147, 170n28, 172n11
Woodward, James, 158n20, 160n23
worship, 42, 67–68, 80, 91, 94

Zygmunt, Joseph F., 152n9, 154n21